THE DOOMSDAY BOOK OF ANIMALS

For my daughter TAROT

THE DOOMSDAY BOOK OF ANIMALS

A Unique Natural History of Three Hundred Vanished Species

by David Day

Foreword by HRH The Duke of Edinburgh KG KT

Illustrated by Tim Bramfitt, Peter Hayman, Mick Loates and Maurice Wilson

EBURY PRESS LONDON

Published by Ebury Press
National Magazine House
72 Broadwick Street
London W1V 2BP

First Impression 1981

Designed by David Cook Associates

Created and produced by
London Editions Limited,
70 Old Compton Street, London W1V 5PA

ISBN 0 85223 183 0

Printed in Great Britain

Acknowledgements

Throughout the two years of research and preparation for this book there have been many people who have helped and to whom I am indebted. First and foremost I owe my thanks to my researcher Seán Virgo for his hard work and enthusiasm in this project.

The British Museum (Natural History) has been a major source for this book and I am grateful to the General Museum Librarians. The Zoology Librarians, Anne Datta, Angela Jackson and Jennifer Jeffrey, were particularly helpful and resourceful, as was Mrs Anne Vale of the Sub-Department of Ornithology at Tring. The Mammals Section also permitted us to use their library.

Special thanks for their involvement as consultants are due to Dr E. N. Arnold of the Reptiles Section and Mr Bishop of the Mammals Section who have most generously advised and helped. Thanks also to Mr Derek Goodwin and Mr Ian Galbraith of the Sub-Department of Ornithology and Mr C. Walker of the Palaeontology Section.

In the United States, Dr David Mech of the US Wildlife Service has been a most able adviser and consultant.

The artists, who have worked hard to research the illustrations, are also indebted to the British Museum (Natural History). The bird illustrators were particularly appreciative of the help that Mr Galbraith and Mr Derek Read at the Sub-Department of Ornithology were prepared to give. Maurice Wilson, the mammals artist, would like to thank the Mammals Section of the British Museum, as well as the Library staff at the London Zoological Society.

At the Booth Museum in Brighton, Mr Adams kindly loaned relevant specimens for illustration, and C. W. Benson at the University Museum of Zoology, Cambridge was most helpful as well.

Finally, I want to thank Rosalie Vicars-Harris of London Editions for her support and her faith in this book and also David Cook, on behalf of myself and the artists, for his wealth of experience and gentle encouragement while commissioning the artwork and designing the book so beautifully.

Contents

Part Two MAMMALS

Part Three REPTILES, AMPHIBIANS AND FISH

APPENDIX

Foreword by HRH THE DUKE OF EDINBURGH

BUCKINGHAM PALACE.

Species have been disappearing from the face of the earth for a very long time. I daresay that if the author had been around about 60 million years ago he could have written an even more horrifying book about the disappearance of the giant saurians and reptiles.

The difference today is that the cause of the disappearance of species is largely if not entirely homo sapiens. Its vast increase in numbers from 450 million to 4,000 million in the last 300 years alone; and its astonishing development of advanced technology is making it increasingly difficult for any of its fellow species to survive.

There can be no finality in the scientific study of endangered and extinct species, but in this comprehensive book, the author sets out what he has been able to discover about the record of extinct species so far. It makes extremely uncomfortable reading but unless a great many people get to know what has happened, what is happening and what is likely to happen in the future in the natural world around us, there will be no chance of preventing further and catastrophic damage.

1981

Introduction

'There is no survivor, there is no future, there is no life to be recreated in this form again. We are looking upon the uttermost finality which can be written, glimpsing the darkness which will not know another ray of light. We are in touch with the reality of extinction.'

Henry B. Hough

The purpose of this book is to put the reader in touch with the reality of extinction. Like the original *Domesday Book*, this is a kind of descriptive census, but it is one that chronicles the last days of vanished creatures and offers a vision of living beauty which has passed.

It is a book that was written out of a deep personal concern. Having grown up on the west coast of Canada, and having spent much of my working life in wilderness regions, my experience of wild animals has been direct and formative. Daily encounters with wildlife through my childhood left their living images imprinted on my imagination. In the forest clearings near my home there were always herds of Black-tail Deer. Everywhere, it seemed, there were the big hunting and scavenging birds; and often, in the harbour, swam otters and seals. There were rare moments as well: my first eye-contact with a mountain lion, and once, alone on a windy point, watching for an hour the astonishing procession of a hundred Killer Whales passing through a narrow strait – their high-keeled backs extending to the far horizon.

Years later, in lumber camps, I would work daily within sight of the elegant, grazing herds of the rare, towering Roosevelt Elk of Vancouver Island. I soon came, with everyone else there, to regard the constant presence of Black Bears with the same offhand familiarity as I had once viewed racoons and squirrels. Other years, working in mainland camps, I would witness the yearly salmon migrations, when the rivers were crowded with spawning fish, and a score or more of Bald Eagles might gather in one place to drag gasping salmon from the shallows and the rapids. Further north again, among the highest mountains on the continent, I was to watch more than a dozen giant Alaskan Brown Bears walking each day with surefooted agility across a mile of high glacier: with careless ease the huge animals leaped over the bottomless blue fissures of ice, as they made their way towards our mining camp.

Now the value of such images as part of a private mythology is obviously very much a personal thing; and the possibility – sometimes, it seems, the probability – of those scenes vanishing involves a sense of personal loss. However, even ignoring for the moment the lives of the animals themselves, the prospect that my child, or her children to come, might never be able to share such experiences is, for me, a tragedy.

As a writer, I feel that the primary, shaping influences of my life came directly out of my encounters with the wilderness. This is particularly true of the years spent in the lumber camps of British Columbia. Writing seemed a means of working out a relationship with the wilderness: of interpreting its hidden language and carrying on a running dialogue with it. My journals of that time came more and more to speak in images drawn from an observation of natural forms which, by careful working and reshaping, might be transformed into an intelligible art form. This was the process by which the powerful visual languages of the North West Indians were distilled into the supreme art form of those tribal peoples, the emblematic totem poles: the huge cedar pillars carved with powerful, archetypal animal images.

In literature, it seemed to me, these images could only be properly conveyed through poetry, and so it was poems that came out of those journals of a wilderness life. Such vitality as my writing has, it owes to the images which emerged from the brooding forest and restless sea.

It owes a great deal, as well, to those men I worked with whose lives were vitally in contact with that same wilderness. Among them were hunters and trappers who are some of the most likeable men I have ever known, as well as perhaps the toughest and most enduring. However, knowing and working with them made me realize that each hunter had unknowingly made a pact with his prey – and that the hunted beast was not, finally, the only victim.

In virtually every professional hunter and trapper I have known, I have encountered extraordinary sentiments and a process of transformation. In the older men, especially, a strange kind of identification comes about. Visions of animals now vanished but once numerous are recalled again and again – waking and dreaming – by those who destroyed them. There are unending stories of the almost mythic size of animals, the vastness of numbers. The stories finally betray a sentiment the men have unknowingly harboured: an admiration and a kind of love for the animals they hunted. Too late, they find in the beasts qualities which they value far more than those of their fellow human beings. I have met cougar and bear hunters who, in old age, have turned bitterly away from mankind, have totally identified with the animals they once hunted, and have come, almost physically, to resemble those animals.

In the two years that have gone into researching and writing this book – largely using the massive archives and libraries of the British Museum (Natural History) – I have come to feel as deeply about many of the strange and beautiful creatures that have been erazed from the Earth's surface, as I do about those that have come out of 'my own' wilderness.

Putting together naturalists' notes, ships' logs and native lore – accounts by those who saw the creatures alive, and often by those who destroyed them – as well as the scholarly monographs and the many general reference works; I have become fascinated with numerous, extraordinary vanished creatures, and saddened, often angered, by their extinction. For in all but a few cases it has been Man through greed or cruelty, carelessness or indifference that has either directly or indirectly been the cause of these extinctions. Yet what moved me most, what drew me into the reality of their vanished existence, were the relics. I have touched the Aurochs' horn, the Huia's long tail-feathers, the pelt of the beautiful Bali Tiger: forever vanished creatures. I have stared at the broken amulet of the last Dodo's skull, and I have peered through the weathered rib cage of a Steller's Sea Cow, into a space large enough to have held that Bali Tiger.

Strangely, one of the most profound of all these experiences was simply holding the huge, polished ivory egg of the Elephant Bird. How extraordinary it was: the largest single cell to exist on Earth in all time – it was a world in itself. I could see how it might be conceived of as the egg of the mythical Roc, and how that egg was the object of many legendary quests. It was the one object that even Aladdin's genie might not win for his master: the repository of all the wisdom of the world.

With the giant egg of this half-ton bird, and later with the tiny form of a vanished Hawaiian Honeycreeper in my hand like a child's soft kid glove, I felt at last in touch with a lost world.

Like the medieval bestiaries which harboured the ancient mythical beasts, *The Doomsday Book of Animals* offers these extinct animals a last resting place. For me, touching the skin of the Thylacine and the horn of the Blue Buck was as miraculous as if they had been the skin and horn of the Griffin and the Unicorn. Like the mythical animals the only life such vanished creatures can now have is in the imagination of Man. To this end *The Doomsday Book of Animals* has been written.

Passenger Pigeon

The Doomsday Book of Animals is about vertebrate extinctions from 1680 to 1980. It is a comprehensive encyclopedia of those extinctions, but because it deals with issues that are no longer the sole concern of the trained natural scientist, it has been written in such a way as to inform, and excite the interest of, the general reader.

The book has been illustrated to show animals as they were in life, not as stuffed specimens. The text presents the facts of their extinction and is often graphically shocking and violent. The juxtaposition is intentional. The message of this book is: 'These animals are extinct. All this beauty and vitality is lost for ever. We can no longer claim to be ignorant of the reasons for their disappearance. Surely the next century need not witness 100 to 1000 more extinctions before the lesson is learned?'

Since life began on this planet three and a half billion years ago, nothing in the history of evolution has approached the sheer destructiveness of the last 300 years. Since the killing of the last Dodo in 1680, there have been at least 300 extinctions of vertebrate animals, more than half of these being full species. Before the expansion of Western Man and his culture, the extinction of an animal species was a rare occurrence. Even during such cataclysmic processes as the 'Great Dying' of the dinosaurs, the rate of the dinosaurs' extinction has been estimated at not greater than one species per thousand years.

It would be quite wrong to use such misunderstood terms as 'natural selection' and 'survival of the fittest' as an explanation for extinction without taking into account the vast scale of millions of years over which evolutionary extinction takes place. To say that the Dodo, Steller's Sea Cow, the Quagga and the Passenger Pigeon became extinct because of evolutionary faults that did not allow them to adapt to new conditions (which Man's technology introduced), is as plausible as explaining the collapse of the Japanese in World War II in terms of genetic flaws: the populations of Hiroshima and Nagasaki could not develop a biological immunity to atomic radiation.

Severe as the devastation of the animal kingdom has been during the last three centuries, this may be just the beginning of a holocaust for the animals. In the listings at the back of this book you will find along with the extinct animals, nearly 400 of the most critically endangered vertebrates – and this is less than one half of the official list of imperilled animals.

For some, this is merely a waiting list. Many of these creatures will undoubtedly become extinct during the next few years. It is estimated that there are today: 1 Abingdon Galapagos Tortoise, 2 Kauai O-O Honeyeaters, 5 Mauritian Ring-necked Parakeets, 5 Javan Tigers, 6 Mauritian Kestrels, 12 Chatham Island Robins, 18 Mauritian Pink Pigeons, about 40 California Condors, about 50 Javan Rhinoceros – and the list goes on.

Even discounting the nuclear threat to our own species, if the human race is blessed with peace and prosperity, a great many animals are doomed. On the huge continent of Africa, the wildlife population has declined by more than 70 per cent since the turn of the century. Further 'development' of nations in Africa and South America, in particular, will spell the end for a great many vertebrates, not to mention the multitude of invertebrates and plants.

In any discussion of this kind, one comes at last to the old assumptions of 'manifest destiny'; to the question: 'Why should we care? What is the use of a wild animal?' There are the larger answers: cogent arguments for the ecological interdependence and balance of all living things. And there is an aesthetic response too, which might run: 'To ask what use there is in a Tasmanian Pouched Wolf or a Hawaiian Honeycreeper, is to ask what use there is in the painting of Rembrandt or the sculpture of Michaelangelo?' Such an argument is unlikely to register with those who do not perceive the inherent moral value and beauty of the thing itself.

However, even on the most elementary and selfish level, there are a good number of solidly logical economic reasons for not over-exploiting wild animal populations. Most obviously, a maximum yield cannot be maintained in a dwindling population. But even further, there can be no logic at all to hunting an animal to extinction. Anyone who hunts an endangered animal for profit, without any quota restriction or 'harvesting' strategy, is acting without any sense of logic, or perhaps in the belief that some divine power will resupply the earth with the animals he has extinguished. The hunter who exterminates his prey destroys his own livelihood.

For me, personally, there is another more basic reason for this concern. To those who are not locked into an anthropocentric view of the world, it is perhaps a more obvious reason than any. The world is not so narrow a place that we can be the sole judges of what is valuable and beautiful in it. Anyone who has spent days and nights watching another animal in the wild will realize that each species perceives a world of its own – totally unlike any other. Time spent with wild creatures makes one realize that there are many conscious life-forms of equal value to our own. If a species becomes extinct, its world will never come into being again. It will vanish for ever like an exploding star. And for this we hold direct responsibility.

As *The Doomsday Book of Animals* is designed to be both a reference book and a consecutive 'read', each chapter and each entry is largely self-contained. There will, therefore, be some minor repetition of geographical locations, ships' voyages and naturalists' identifications, for example.

The order of the chapters, and the specific groupings within them, have been chosen in line with popular, rather than strict taxonomic progression. Sometimes it has been more logical to follow geographic rather than generic extinction sequences. For those who wish to follow taxonomic classification there is such a listing, of both extinct and endangered animals, in the appendix.

For easy reference there are marginal keys beside the main text giving the common and scientific names of each species, the date of extinction and habitat location. Nearly all extinct animals have been illustrated. Notable exceptions are in the 'Giants' and 'Bats and Rats' chapters; the reason being that many of these animals (though extinct within historic times) are known only from bones. A further exception is the Reptiles chapter, where new information and recent research publications (1981) on the reptiles of the West Indies came rather too late for illustration.

The date of extinction is generally given as the date of the last specimen taken, although in some cases it may be the last verifiable sighting. The limit of 300 years (1680 to 1980) has been generally observed, but in a few cases – such as the Aurochs and some of the Moas which slightly preceded the Dodo's extinction – exceptions were made to give a fuller perspective of Man-related extinctions.

A note concerning metric conversion: converting from imperial to metric measurements may sometimes give a misleading impression of precision. To say that a bird recorded in 1870 as being 4½in. long, measured 11·43cm, would falsify both the nineteenth-century naturalist's claims and our present state of knowledge. Animal skins in museums can stretch or shrink, so the original measurements must be given, yet they were often made with a tape-measure in the field. Thus the metric equivalent has been rounded off, and must be taken as approximate only. This applies also to conversions of miles, square miles, acres and altitudes measured to the nearest 500 feet.

This book has dealt with only the most complex life forms, the vertebrates. However, it is difficult to discuss vertebrate animals without considering aspects of the ecological system which supported them. In many cases, the destruction of plant systems on a large scale resulted in the disappearance of the animals who depended on them. Therefore, to draw attention to the complexity of the issues involved in ecological preservation today, there is a short essay on endangered and extinct plants in the appendix section.

The appendix also has an alphabetical index of common names, scientific names, places and people; as well as a select bibliography. An extremely important part of the appendix is the 'Atlas of Extinction' which will help orient the reader in some of the more remote areas to which his reading will take him in the course of this book.

Part One

BIRDS

Elephant Bird *Aepyornis maximus* Extinct *c.* 1700

The Giants

ELEPHANT BIRDS, MOAS, OSTRICHES AND EMUS

ELEPHANT BIRD
Aepyornithidae
Aepyornis maximus
EXTINCT *c.* 1700
Madagascar

In the year 1298, while languishing in the prison of Genoa, the Venetian merchant Marco Polo committed to paper his memoirs of 26 extraordinary years of travel through the Orient. In chapter 33, *Concerning the Island of Madagascar,* the Venetian recorded strange accounts of giant birds, and that 'the Great Khan sent to those parts to inquire about these curious matters'. The tale of the envoys of the Khan, of course, reflected one of the many Arabian stories of the giant Roc, such as those which relate to the adventurer-heroes, Sinbad and Aladdin. Yet had the envoys of the Khan made their journey to Madagascar they indeed would have brought back news of a Roc, and relics that were in little need of exaggeration. By Marco Polo's time Saracen and Indian traders had plied the ocean along the African coast for many centuries, and knew the giant birds, not as mythical beings at all, but as a living if startling reality.

Tales of this Roc-bird of Madagascar can be traced back to the days of Carthage; although often the shape, size and powers of the giant bird were changed or enlarged in the telling. Still, it was not until the sixteenth century that there was any direct European

contact. Dutch, Portuguese and French seamen returned from the Indian Ocean with huge eggs taken as curios. These eggs astounded the few educated men who saw them.

They were as much as 91cm (3ft) in circumference and had a fluid capacity of over two imperial gallons (9 litres) – the equivalent of 200 domestic chicken eggs. They were three times the size of the eggs of the largest dinosaurs. What bird could lay such eggs?

It came to be called the Elephant Bird *(Aepyornis maximus)* and it lived in the primeval wilderness of that 1,600km (1,000 miles) long island, Madagascar. To the first men who encountered it, the Elephant Bird must have been a terrifying apparition as it loomed out of the dense tropical forest or strode across the dune-sands of the shoreline. It was 300cm (10ft) tall, weighed over 504kg (1,100lb) and was the largest bird that ever lived on earth. It was a ratite, a giant running bird. It had massive legs armed with taloned claws. Its huge body was covered in strange, bristling, hair-like feathers similar to the Emus'. It had vestigial wings and a powerful, serpent-like neck with a head and beak like a broad-headed spear. Although it did not fly and feed on elephants, like the creature of Sinbad's tale, it was scarcely less amazing than the Roc of Marco Polo's account.

The Elephant Birds were born to an age when the birds were the dominant life form on earth, and on Madagascar for perhaps sixty million years they remained lords of their world. Though they were herbivores, they had little to fear. They were protected like the elephant itself by size. If need be there were the slashing taloned feet and a blow like a heavy spear thrust from the beaked head. There was no predator that could hope to threaten them; none, that is, until the advanced hominid called Man entered their world.

These giant birds had through their long history adapted to many changes and could be numbered among the most successful of bird species – enduring more than thirty times as long as humans have existed at all. Indeed, contact with humans is not likely to have been more than a few thousand years in duration: the direct result being the total eclipsing of the giant species.

Aepyornis maximus, however, is known not to have been the only ratite on Madagascar. From fossil evidence, it is likely that there were between three and seven species of Elephant Bird or Aepyornithidae, varying in height from 300cm (10ft) to less than 90cm (3ft). However, nearly all these were prehistoric forms, most dying out before *Homo sapiens* had even evolved. Nonetheless, it seems likely that at least one other, smaller ratite – probably the form called *Mullerornis* – survived into historic times with the giant Elephant Bird.

Because there are no reliable historical records of the pre-European history of Madagascar, it is difficult to get a clear picture of the reasons for the Elephant Birds' extinction. However, it is likely that they were hunted by primitive men for one or perhaps two thousand years before European contact. This hunting would probably not have endangered the species to so great an extent had it not been combined with egg collecting. It was probably this form of predation more than anything else that so reduced the birds' numbers, although it must have taken considerable habitat destruction to extinguish the species finally.

The Malagasy people had had contact with Arab traders over several centuries, but they fiercely resisted colonization. Dutch and French expeditions established coastal settlements after 1509, but the heavily populated interior was closed to them for another 150 years. Of early Arab and European influences, again, there is little in the way of records. Merchants and men involved in such trades as slave-running and pirating were not the type to keep ornithological notes on exotic birds. Still, it cannot be doubted that trade and skirmishing battles resulted in the importation of steel weapons and tools – and even muskets in later times. Against guns and the devastation of the forests by burning, cutting and clearing, the already critically dwindled population of Elephant Birds retreated, driven to ever remoter regions.

Evidently by 1658 the giant birds had already withdrawn from the major part of their habitat. In his report of that year the Sieur Etienne de Flacourt, Director of the French East India Company and Governor of Madagascar, wrote of the Elephant Bird under its local name: 'The Vouron Patra is a giant bird that lives in the country of the Amphatres people [in the south of Madagascar], and lays eggs like the Ostrich; so that people of these places may not catch it, it seeks the loneliest places.'

On an island – even one as large as Madagascar – there proved to be no place lonely enough for the Vouron Patra. When the final blow fell – whether by gun, axe or fire – is not known, although by 1700 the Elephant Bird was almost certainly extinct. Thus, almost before the western world had learned of its existence, the largest bird that ever lived had vanished.

In the mythology of the Maoris of New Zealand there is an equivalent to the giant Roc of the Arabic world. This is the 'Poua-Kai', a huge bird of terrific size and strength which, in a great battle, destroyed half the warriors of a powerful tribe with its terrible rending talons and thrusting beak.

It is not strange that the Maoris should have such tales in their mythology. They were a Polynesian people who did not arrive in New Zealand until the tenth century AD. What greeted them was an astonishing world that seemed frozen in time. Far more isolated than Madagascar, New Zealand showed how the world had fared over sixty million years before: it was a semi-tropical realm dominated by birds. Indeed, there were no land mammals except bats.

The dominant life-forms were the giant land birds which the Maoris called 'Moas'. These were huge ratite 'running birds' like the Elephant Bird, but they inhabited the grasslands and forest-fringe in extraordinary numbers and variety. Scientists later gave them the family name Dinornithidae, 'terrible birds'. The largest of the Moas was twice the height and three times the weight of a large man. It was easily the second largest of all the earth's birds, for thought at 275kg (600lb) it was only about half the weight of the Elephant Bird, it was even taller, and reached a height of 4m (13ft).

The Polynesian invaders were an active and aggressive people, and they soon became a 'Moa-hunting culture'. For the first time in perhaps one hundred million years there were predators on New Zealand that were a threat to these birds. The impact was nothing

MOAS

Dinornithidae
Dinornis maximus
Dinornis robustus
Dinornis gazella
Pachyornis septentrionalis
Pachyornis elephantopus
Pachyornis mappini
Emeus huttoni
Euryapteryx gravis
Euryapteryx geranoides
Anomalopteryx parvus
Anomalopteryx didiformes
Anomalopteryx oweni
Megalapteryx didinus
Megalapteryx hectori
Megalapteryx benhami
EXTINCT 1500–1850
New Zealand

less than catastrophic for the Moas.

There may have been as many as 25 species of Moa in the Dinornithidae family which evolved over 100 million years. The earliest forms were quite small, but many giant forms soon developed. Of these, it seems as many as 15 species were contemporary with Man, although some were probably quite rare by the time of the Maori invasion.

Of those Moas that came into contact with Man, there were six genera, the largest being the Dinornis with the species *Dinornis maximus, Dinornis robustus* and *Dinornis gazella. Dinornis maximus* was the 4m (13ft) giant form and may have survived longest of all the Moas; certainly evidence and sightings suggest it probably only became extinct around 1850. Other huge forms of Moa were the shorter, but very heavy 'Elephant-Footed Moas' of the genera Pachyornis and Emeus. The species *Pachyornis septentrionalis, Pachyornis elephantopus* and *Pachyornis mappini*, together with *Emeus huttoni*, were evidently hunted to extinction by the Maoris between 1100 and 1500. The medium-sized, though powerfully built, Moas of the genus Euryapteryx – with its two species *Euryapteryx gravis* and *Euryapteryx geranoides* – probably survived as late as 1700. There were also six species of so-called 'Pygmy Moas' which stood between 90cm and 120cm (3–4ft) in height. These were the species *Anomalopteryx parvus, Anomalopteryx didiformes* and *Anomalopteryx oweni;* and *Megalapteryx didinus, Megalapteryx hectori* and *Megalapteryx benhami*. Nearly all of these seem to have died out through Maori and European agency by 1800, though there is evidence, and one very convincing sighting record, to suggest that one form of 'Pygmy Moa' (possibly *Megalapteryx didinus)* may have survived into this century. Some believe there is a slim chance that these Moas may still survive in the wilderness of Fiordland.

News of these giant birds was slow in reaching Europe. It was not really until 1838, when an Englishman named John Rule brought the fragment of a huge leg-bone out of New Zealand to the famous palaeontologist Richard Owen in London, that any notice was taken of the Moas in scientific circles. Even so, it took several more years and an entire crate load of bones to convince natural scientists of the Moas' existence.

The consignment of bones had been sent by the Buckland geologist and missionary, Revd William Williams, in 1843. Williams had independently made a study of the birds, and in 1842 recorded a sighting by two English whalers near Cloudy Bay, in Cook Straits. Of the incident Williams wrote: '. . . the natives there had mentioned to an Englishman of a whaling party that there was a bird of extraordinary size to be seen only at night on the side of a hill near there; and that he, with the native and a second Englishman, went to the spot; that after waiting some time they saw the creature at some little distance, which they describe as being fourteen or sixteen feet high. One of the men proposed to go nearer and shoot, but his companion was so exceedingly terrified, or perhaps both of them, that they were satisfied with looking at him, when in a little time he took alarm and strode up the mountain.'

When 'discovery' of these giant birds created something of a sensation in popular and scientific circles, Queen Victoria asked that a special report be written, for her own edification, on the subject.

Giant Moa *Dinornis maximus* Extinct *c.* 1850

Within a few decades huge bone finds of a thousand or more birds had been made in dried swamplands, caves and Maori middens.

Besides the extraordinary size of the Moas, one of their most striking features was that alone of all the world's birds, they were totally without humeri (upper arm-bones) and thus did not have a trace of even a vestigial wing-structure.

As interest was aroused, other accounts were brought to light. A New Zealand resident, John White, interviewed a number of sealers during the 1850s who claimed to have eaten Moas on the South Island. But the most vivid and detailed account of the giant birds was given to him by an old Maori, also on South Island, who described the Moas' appearance, habitat, feeding and nesting habits, as well as the traditional manner of hunting and cooking them. He recalled the fierce, booming male Moas, guarding the females at their nests, and how the Maoris would ambush them on their trails, spearing them in the sides so that, as they ran, the spears would lash against the scrub bushes and exhaust them.

A different hunting method was described in the early nineteenth century by the Maori, Kawana Papai, who said he had taken part in hunts when he was a boy (around 1790). He said that they rounded up the Moas and speared them; that a Moa kicked viciously but, of course, had to stand on one leg to kick; and that the hunters then struck at its standing leg and thus brought it down.

As with the Elephant Bird, a great many huge eggs were gathered, some measuring as much as 25cm (10in.) in length. And as late as 1865, a Moa egg was discovered near Cromwell, which still contained an embryo of a Moa chick. One of the largest eggs was found in a Maori gravesite at Kaikura. An old Maori chieftain was buried in a sitting position with the huge egg of an extinct Moa grasped in the skeletal bones of his hands.

There was something strongly presentient in that figure in the Maori grave. We might recall how the Maoris, when faced with the destruction of their civilization by war and disease, used in their poetic lament the proverbial expression: 'Ka ngaro i te ngaro a te Moa' – Alas, we are 'Lost as the Moa is lost'.

ARABIAN OSTRICH

Struthiornidae
Struthio camelus syriacus
EXTINCT *c.* 1941
Syria and Arabia

Today the largest living bird is, of course, the Ostrich. This ratite reaches maximum weights of over 114kg (250lb). It is now found in the wild only in Africa, though until the mid-twentieth century, there was a distinct subspecies of Ostrich in the Middle East.

Rather smaller than the African forms, the Arabian Ostrich *(Struthio camelus syriacus)* was common in Syria and Arabia until 1914. World War I brought a great influx of modern firearms into the Middle East, and the realization that a considerable amount of money could be made supplying the European fashion for Ostrich feathers. Thus the 'Na-ama' (as the Arabs called the Ostrich) was ruthlessly hunted for its plumes and, with the coming of the automobile, was even pursued and shot for 'sport'.

Because of the ready market for this bird and the poverty of the local peoples, the extermination of the Na-ama was extremely rapid. It was seldom seen after the mid-thirties anywhere in its former range: the desert regions from the Euphrates in the east, to Sinai in the west, to southern Saudi Arabia in the south. Despite reports of a bird drowned in a flash flood in 1966, the last authenticated Na-ama was shot in Bahrein in 1941.

TASMANIAN EMU **DWARF EMU**
Dromaiidae *Dromaius novaehollandiae diemenensis* EXTINCT *c.* 1850 Tasmania *Dromaius novaehollandiae diemenianus* EXTINCT *c.* 1830 Kangaroo Island

The 'Ostrich' of Australia is the Emu, the next largest of living birds. Like the Ostrich, the Emu has suffered (and still suffers) considerable depradation from Man. During the 1920s, 30s and 40s the Australian government paid a bounty for Emu beaks and eggs. In 1932, in a bravado act that was more symbolic than effective, federal soldiers were sent out to use machine guns against the birds.

Although Emus survive on the continent of Australia, they have been extinct for over a century on Tasmania and the smaller Kangaroo Island, where distinct subspecies were once common.

The Tasmanian Emu *(Dromaius novaehollandiae diemenensis)* was smaller than the Australian bird and had no black feathers on its foreneck. It survived no later than 1850. The Kangaroo Island subspecies, the Dwarf Emu *(Dromaius novaehollandiae diemenianus)* – the smallest and darkest of all Emus – was extinct twenty years earlier in 1830. Both these island subspecies were destroyed by hunters and man-started bush fires.

The convict settlements of Tasmania proved devastating to all indigenous Tasmanian life – both animal and human. And on the small, 144km by 56km (90 mile by 35 mile) Kangaroo Island, extinction was achieved with extraordinary rapidity. The island was only discovered in 1802 and whalers and seal hunters were the only inhabitants, but in less than three decades the birds were hunted out. By the time of the first settlement by the English in 1836, the once numerous Emus had been extinct for several years.

Moa
Dinornis maximus
Extinct *c.* 1850

Moa
Pachyornis septentrionalis
Extinct *c.* 1500

Moa
Emeus huttoni
Extinct *c.* 1500

Moa
Megalapteryx didinus
Extinct *c.* 1850?

Common Dodo *Raphus cucullatus* Extinct *c.* 1680

Gentle Doves

THE PIGEONS AND THEIR ALLIES

DODO
Raphidae *Raphus cucullatus* EXTINCT *c.* 1680 Mauritius

Every schoolchild knows the Dodo, the butt of so many jokes about stupidity and obsolescence. The fate of the peculiar bird that once lived on the island of Mauritius some 800km (500 miles) west of Madagascar, is known to all. The Dodo is as sadly famous for its extinction, as the mythical Phoenix is for its resurrection.

The first explorers of Mauritius variously labelled the Dodo: a wild turkey, a cassowary, a 'hooded swan', a booby and a 'bastard ostrich'. In fact, the Dodo was a large, flightless dove which, primarily because of its gentle dove-like qualities, became extinct in 1680.

The Dodo *(Raphus cucullatus)* was without doubt the largest and strangest dove ever to have lived. Its appearance fascinated the first European observers. Among the earliest accounts of the Dodo is one left by the Englishman Sir Thomas Herbert, who visited Mauritius in 1627: 'First, here and here only . . . is generated the Dodo, which for shape and rareness may antagonize the Phoenix of Arabia: her body is round and fat, few weigh lesse then fifty pound, are reputed of more for wonder then for food, greasie stomackes may seeke after them, but to the delicate, they are offensiue and of no nourishment.

'Her visage darts forth melancholy, as sensible of Nature's injurie in framing so great a body to be guided with complementall wings, so small and impotent, that they serue only to prove her bird.

'The halfe of her head is naked seeming couered with a fine vaile, her bill is crooked downwards, in midst is the thrill, from which part to the end tis of a light greene, mixt with pale yellow tincture; her eyes are small and like to Diamonds, round and rowling; her clothing downy feathers, her traine three small plumes, short and inproportionable, her legs suting to her body, her pounces sharpe, her appetite strong and greedy.'

Although Mauritius had undoubtedly been visited by Arab traders before the Portuguese 'discovered' the Mascarene islands in the early sixteenth century, no human settlements had been established. So, until the Portuguese came, Mauritius remained much as it had been for tens of millions of years. Like so many remote island groups the Mascarenes had no mammals, but were dominated by birds. As an early observer noted: 'The birds (of which the island is full) are of all kinds: Doves, Parrots, Indian Crows, Sparrows, Hawks, Thrushes, Owls, Swallows, and many small birds; white and black Herons, Geese, Ducks, Dodos . . .' The birds adapted to take maximum advantage of their habitat and evolved into many unique, strange and colourful species. When the European ships arrived this unique collection of birds provided the crews with easy food and sport, for the birds, with no experience of predators, would often walk right up to their destroyers. One can imagine the novelty these birds presented to the sailors who, confined to meagre rations for many weeks and months, must have relished the thought of so much fresh, available meat and easy sport.

The Portuguese were followed by the Dutch in 1598, when Cornelius Van Neck made Mauritius a Dutch possession. After that time many ships stopped there on long journeys across the Indian Ocean and indulged in countless pathetic slaughters of the unwary birds. Another Dutch captain reported: 'We lived on Tortoises, Dodos, (Pink and Blue) Pigeons, . . . Grey Parrots and other game, which the crews caught by hand.' None of these animals are now found on the island, except the Mauritian Pink Pigeon, of which there are less than twenty individuals in the world today.

Despite the slaughter of wildlife carried out by the hundreds of European ships that visited Mauritius, the Dodo survived. In the more inaccessible parts of the island breeding colonies remained elusive enough to escape marauding seamen. But in 1644, the Dodo's fate was sealed, for the island became a Dutch colony. The colonists seemed to display a grim dedication to the cause of exterminating the Dodo, although the bird was neither a nuisance nor a menace and, according to most accounts, its flesh was very poor eating, being tough and rather bitter. There must have been little sport in walking up to an almost tame bird and crushing its skull with a wooden club. As the birds could not fly, and any able-bodied man could run faster, the outcome of any hunt was a foregone conclusion.

Beyond the outright slaughter of Dodos carried out by men, other equally devastating enemies accompanied the colonists. For the settlers brought with them not only their pet dogs and cats, but also monkeys, swine and the inevitable rat. These creatures readily adapted to the island's wilderness and rapidly cleared it of Dodos, even in the areas most inaccessible to men. Roaming dogs were most effective at destroying the adult birds, while cats, rats, monkeys and pigs quickly rooted out both eggs and chicks. Ironically, in this tropical land the wild domestic animals multiplied at such a rate, they became the nuisance that the birds had never been. At one point a plague of rats was so severe that it caused the 'Hollanders' on the island to emigrate. The destruction caused by feral pigs to the forests and farms became so great that a hunt was organized in which 1500 were killed in a day. That same year a chronicler also mentioned a single garden in which 4000 monkeys frolicked. By 1680, the island was completely overrun by men and the animals they had introduced, and that unique creature called the Dodo was gone forever.

RODRIGUEZ SOLITAIRE

Raphidae
Pezohaps solitarius
EXTINCT c. 1780
Rodriguez

The Dodo of Mauritius was not the only species of giant dove in the family Raphidae: there were three others. Two of these were the Dodo's elegant cousins, the Solitaires; one of which lived on the island of Réunion 160km (100 miles) to the south west of Mauritius, while the other inhabited Rodriguez some 480km (300 miles) to the east.

According to the journals of all who saw the Solitaires, they were accounted both delightfully beautiful and delightfully edible. Indeed, the earliest observer of the Solitaire of Rodriguez *(Pezohaps solitarius)* seemed almost enraptured in his description of the birds.

'The Females are wonderfully beautiful, some fair, some brown; I call them fair, because they are of the colour of fair Hair. They have a sort of Peak, like a Widow's upon their Beak, which is of a

Rodriguez Solitaire *Pezohaps solitarius* Extinct *c.* 1780

Réunion Solitaire
Ornithaptera solitarius

dun colour. No one feather is straggling from the other . . . The Feathers on their Thighs are round like shells at one end, and being there very thick, have an agreeable effect. They have two Risings on the Craws, and the Feathers are whiter there than the rest, which lively represents the fine bosom of a Beautiful Woman. They walk with so much stateliness and good Grace, that one cannot help admiring and loving them.'

The writer, François Leguat, was a remarkable man – a Huguenot refugee from France who spent two years of his exile on the previously uninhabited Rodriguez. His descriptions of the Solitaires, published in his 1708 memoirs, were considered fanciful for two centuries until archaeology proved that so much of what he said was true that the rest must be taken seriously. He describes their mating dances and displays, when they whirred their short wings and clapped them 'like a rattle' against their sides. He tells how they guarded their single chicks, driving off all other Solitaires from their territory, and how they mated for life – 'these two Companions never disunite'. But his most remarkable description is of their annual 'betrothal' ceremonies: 'Some days after the young one leaves the nest a Company of 30 or 40 brings another young one to it; and the new-fledged Bird with its Father and Mother joyning with the Band, march to some bye Place. We frequently followed them, and found that afterwards the old ones went each their way alone, or in Couples, and left the two young ones together, which we call a *Marriage*.'

Leguat was not the only traveller moved to talk of this graceful long-necked bird in human terms: 30 years later another wrote: 'When caught they make no sound, but shed tears', and said that they pined away quickly, refusing to eat in captivity.

Unlike their portly cousin from Mauritius, the Solitaires could be aggressive: they pecked fiercely and their apparently useless wings had knobs 'the size of musket balls' at their tips with which they could strike as well as making that rattling sound. All in all they seem to have earned more respect from their human visitors than their relatives on the other islands.

Yet, the Solitaire – despite its 'usefulness', 'intelligence' and 'beauty' – soon became quite as extinct as its 'useless', 'stupid' and 'ugly' cousin, the Dodo of Mauritius. Though it was still reportedly alive in 1761, a long-time inhabitant of Rodriguez in 1831 had never seen or heard of the Solitaire. It was certainly gone by 1800.

RÉUNION SOLITAIRE

Raphidae
Ornithaptera solitarius
EXTINCT *c.* 1700
Réunion

The Réunion Solitaire *(Ornithaptera solitarius)* seems to have been a slightly larger and more sturdily built bird than the Rodriguez species. Its legs were longer and muscled as if for running at greater speeds. It also had a large tuft-like feathered tail, while its cousin had practically no tail at all. Like the Rodriguez species it was extensively hunted for food, and, probably because it was a much better tasting bird, it became extinct about 1700, a good 70 years before its relative on Réunion, the White Dodo.

Both Solitaires might have been forgotten altogether, dismissed as fairy-tale birds, had not several skeletons on Rodriguez been unearthed in 1875 and later, and proved that Leguat's descriptions of the birds, including the widow's peak and the 'musket-ball' wing-clappers, had been in every way accurate.

No such evidence exists to testify to the existence of the fourth Didine species: the White Dodo of Réunion *(Victoriornis imperialis).* Yet two vivid eye-witness accounts and a number of paintings, one of high quality, have persuaded scientists that the bird did exist and differed in significant ways from the Mauritian Dodo.

In an account of a voyage by Captain Carleton in 1613, J. Tatton wrote of Réunion: 'There is a store of Land-fowl, both small and great, plentie of Doves, great Parrots and such like; and a great fowl of the bigness of a turkey, very fat, and so short-winged that they cannot flie, being white, and in a manner tame; and so are all other fowles, as having not been troubled or feared with shot.' Needless to say this state of affairs was short-lived. This Dodo was obviously as defenceless as its Mauritian counterpart – witness this 1646 description by a Dutch traveller: 'There were also some Dod-eersen which had small wings, but could not fly; they were so fat that they could scarcely walk, for when they trod their belly dragged along the ground.' It should be pointed out that the Dodos' waistlines varied with the season (some authorities even believe that they shed their horny bill-sheaths once a year as part of the general moult and wasting) but the picture, once again, is one of defencelessness.

How long the White Dodo survived can only be guesswork: possibly as late as 1770. Only the greater size of Réunion, and the slower growth of settlement seems to have allowed it to outlive the Mauritian Dodo so long. It is not impossible that the very last White Dodo died in Europe, for the Governor of Mauritius and Réunion between 1735–46, de la Bourdannaye, sent one to the Directors of the Company 'as a curiosity'. It must have been one of the last, if it survived the journey.

The accurate, or at any rate convincing aquarelle paintings by Pieter Withoos in 1680 show that at least one White Dodo *did* reach Europe alive and he pictures it in a park among common European waterfowl which are accurately represented.

He portrays the Dodo with buttercup-yellow wings, a high-arched tail and a faint bluish tint towards the end of its otherwise white plumage. The eyelid was bright red, the legs and feet were ochre-yellow, with black nails (proving that it was not an albino), and the beak far less hooked and pointed than the Mauritian Dodo's. Ungainly it may have been, but in its tropical native setting it must have looked spectacular.

If the curiosity collectors who brought both Dodos for exhibit in Europe had had any of the sense of mission shown by the better modern zoos, these giant doves might today be at large in our parks, however doomed their native habitat. A creature which could survive the sea voyages of that period was tough – it could have survived and it probably could have bred.

One of the Dodo family was on show in the London of Charles I when, in 1638, Sir Hamon Lestrange saw its poster picture on display and went to look. He saw 'a great fowl, somewhat bigger than the largest Turkey Cock . . . but stouter and thicker and of a more erect state . . . The keeper called it a Dodo, and in the end of a chymney in the chambers there lay a heape of large pebble stones, whereof hee gave it many in our sight, some as big as nutmegs and the keeper told us she eats them (conducing to digestion) . . .'

WHITE DODO

Raphidae
Victoriornis imperialis
EXTINCT *c.* 1770
Réunion

White Dodo
Victoriornis imperialis

Members of the Dodo family certainly did eat stones – they held them in their gizzards to crush food; 'bezoars' are invariably found with the skeletons of Mauritian Dodos and Solitaires and were, for a time, valued for their 'medicinal' properties. Nevertheless, the thought of the bird being fed stones in a dark London house, 6000 miles from the sunny Mascarenes, is a depressing one.

PASSENGER PIGEON

Columbidae
Ectopistes migratorius
EXTINCT *c.* 1914
Eastern North America

After the Dodo, the Passenger Pigeon *(Ectopistes migratorius)* is certainly the most famous case of animal extermination. It is also the most astonishing and unbelievable of all extinctions.

As the Elephant Bird of Madagascar was without doubt the largest bird ever to live, the Passenger Pigeon or Migrating Dove was the most numerous. As late as 1860 any naturalist or layman might easily have argued that the Passenger Pigeon was, in biological terms, the most successful species of bird on earth. Its numbers were so great, its territories so vast, and its strong body so well designed for its needs and habitat, that it is almost incredible that it could have been exterminated within the short space of 50 years.

No one knows how many Passenger Pigeons there were, but it has been estimated that they accounted for nearly 40 per cent of the entire bird population of North America. In 1870, when their numbers were considerably diminished by relentless hunting, a single flock 1·6km (one mile) wide and 510km (320 miles) long, containing not less than 2,000 million birds, passed over Cincinatti on the Ohio River.

The noted American ornithologist Alexander Wilson went in 1806 to a Passenger Pigeon breeding ground in Kentucky. It was 64km (40 miles) in length and several miles wide. The ground was scattered with the broken limbs of trees, dung, masses of eggs, and squab chicks which were being fed upon by herds of hogs. The trees each held more than a hundred nests. He wrote of a great flight of these birds going to a nesting site of equal size some 96km (60 miles) away. The Pigeons were travelling at the speed of a mile a minute and high beyond the reach of gunshot, very close together and many layers deep: '. . . as far as the eye could reach, the breadth of this vast procession extended; seeming everywhere equally crowded.' He calculated the number of birds to be 2,230,272,000. On the modest assumption that each bird consumed half a pint of nuts and seeds a day, he estimated this flock would consume 17,424,000 bushels each day.

Similarly, in the autumn of 1813, that most famous of ornithologists and illustrators, James Audubon, was travelling in a wagon from his home on the Ohio River to Louisville, Kentucky, some 88km (55 miles) away, when a column of Passenger Pigeons filled the sky so the 'light of noonday sun was obscured as by an eclipse'. At sunset he reached Louisville and still the birds passed over in a solid mass. For three days other flocks followed this first one. Audubon calculated the size of only one of the columns of birds. He estimated that a flock of birds one mile wide passing overhead for three hours at a mile a minute would equal 1,015,036,000 creatures. He also estimated that this flock would consume 8,712,000 bushels of feed a day.

Like Wilson, Audubon went to roosting and nesting sites. He saw huge broken limbs of trees and many trees, themselves more than 60cm (2ft) in diameter, broken off a few feet from the ground simply by the weight of the birds. The dung was so deep on the forest floor

Passenger Pigeon *Ectopistes migratorius* Extinct *c.* 1914

he mistook it for snow at first. When the flock returned there was unbelievable confusion and uproar; the wings of so many birds were like a gale and the sound of their landing was like thunder.

The Passenger Pigeon's habitat seems to have been virtually the whole of forested North America, although its primary territory could be said to have been east of the Mississippi from the Hudson Bay to the Gulf of Mexico. The pigeons seemed to have fed primarily on the acorns and nuts of the hardwood forests, as well as the seeds of coniferous woodlands. They also fed on grains, grass seeds, all manner of berries and some insects.

The great migrations of these birds seemed to have been related to food gathering conditions as well as breeding habits. When food was plentiful in the woodlands, they came together. When it was scarce, they dispersed into small flocks. Passenger Pigeons were strong and swift flyers. They maintained constant speeds of more than 96km (60 miles) per hour, and were capable of flying 1600km (1000 miles) in a day. Thus, despite their numbers, they seemed always capable of finding sufficient food because of their ability to range so widely in a matter of days.

In 1917, the Massachusetts State Ornithologist, Edward Forbush, recorded what he believed to be the best authentic description of the nesting habits of the Passenger Pigeon. It is an extraordinarily erudite account by a Chief Pokagon, a full-blooded American Indian, said to be the last Pottawottomi chief of the Pokagon band.

'About the middle of May, 1850, while in the fur trade, I was camping on the head waters of the Manistee River in Michigan. One morning on leaving my wigwam I was startled by hearing a gurgling, rumbling sound, as though an army of horses laden with sleigh bells was advancing through the deep forests toward me. As I listened more intently, I concluded that instead of the tramping of horses it was distant thunder; and yet the morning was clear, calm and beautiful. Nearer and nearer came the strange commingling of sleigh bells, mixed with the rumbling of an approaching storm. While I gazed in wonder and astonishment, I beheld moving toward me in an unbroken front millions of pigeons, the first I had seen that season. They passed like a cloud through the branches of the high trees, through the underbush and over the ground, apparently overturning every leaf. Statue-like I stood, half-concealed by cedar boughs. They fluttered all about me, lighting on my head and shoulders; gently I caught two in my hands and carefully concealed them under my blanket.

'I now began to realize they were mating, preparatory to nesting. It was an event which I had long hoped to witness; so I sat down and carefully watched their movements, amid the greatest tumult. I tried to understand their strange language, and why they all chatted in concert. In the course of the day the great onmoving mass passed by me, but the trees were still filled with them sitting in pairs in convenient crotches of the limbs, now and then gently fluttering their half-spread wings and uttering to their mates those strange, bell-like wooing notes which I had mistaken for the ringing of bells in the distance.

'On the third day after, this chattering ceased and all were busy carrying sticks with which they were building in the same crotches of the limbs they had occupied in pairs the day before. On the morning of the fourth day their nests were finished and eggs laid. The hen birds occupied the nests in the morning, while the male birds went out into the surrounding country to feed, returning about 10 o'clock, taking the nests, while the hens went out to feed, returning about 3 o'clock. Again changing nests, the male birds went out the second time to feed, returning at sundown. The same routine was pursued each day until the young ones were hatched and nearly half-grown, at which time all the parent birds left the brooding grounds about daylight. On the morning of the eleventh day after the eggs were laid, I found the nesting grounds strewn with egg shells, convincing me that the young were hatched. In 13 days more the parent birds left their young to shift for themselves, flying to the east about 60 miles, when they again nested. The female lays but one egg during the same nesting.

'Both sexes secrete in their crops milk or curd with which they feed their young until they are ready to fly, when they stuff them with mast and such other raw material as they themselves eat until their crops exceed their bodies in size, giving to them an appearance of two birds with one head. Within two days after the stuffing they become a mass of fat – 'a squab'. At this period the parent bird drives them from the nests to take care of themselves, while they fly off within a day or two, sometimes hundreds of miles, and again nest.

'It has been well established that these birds look after and take care of all orphan squabs whose parents have been killed or are missing. These birds are long-lived, having been known to live 25 years caged.'

Passenger Pigeons *Ectopistes migratorius*

It really does seem inconceivable that a bird as numerous as the Passenger Pigeon could have been so rapidly exterminated. Yet the determination of the men who destroyed them was as extraordinary as the numbers of their quarry. In the 1850s a single New York merchant reported he was selling 18,000 pigeons a day. There were many others like him in New York, as well as Chicago, St Louis, New Orleans, Boston, and virtually every city and town in northern and eastern America.

The demand for these birds was phenomenal. Adult birds were the cheapest meat that could be bought; the squabs were a tender delicacy. The gizzards, entrails, blood and even dung were marketed as medical cures for gallstones, stomach aches, dysentery, colic, infected eyes, fever and epilepsy. Pigeon down and feathers were used for pillows and quilts. There was also a large market for live birds. Sportsmen who indulged in trapshooting bought up perhaps a million birds a year. A shooting club for a week's competition might bring in 50,000 birds, nearly all of which would die either by being shot or having their wings or necks broken by being hurled from the catapult traps. One sporting gentleman might kill 500 or more of these birds in a day's shooting.

After 1860 pigeon hunting became a full time occupation for several thousand men. With the advent of the telegraph and the railroad, hunters were able to follow and slaughter the migrating birds wherever they landed. From then on nesting grounds were seldom safe. The birds were searched out, harried and destroyed. Hundreds of railway boxcars were sent with the hunters and waited to be filled with the carcasses. Local part-time hunters generally used guns, clubs, poles, smudge pots and even fire to kill the birds at nesting sites. The best professional hunters used huge, specially designed traps and nets. Some of the large nets were baited with decoy birds. These birds were called 'stool pigeons'. They were captured birds with their eyes sewn up and their legs pinned to a post or 'stool'. The fluttering wings of these blind birds attracted other pigeons which were then caught up in the huge nets and slaughtered. Some of the net traps were capable of capturing 2000 birds at once.

The pressure of the hunters was relentless. As late as 1878, near Petoskey in Michigan, hunters descended on a nesting sight 64km (40 miles) long and 5 to 16km (3 to 10 miles) wide. Their efficiency was astounding. In this one hunt it is estimated they destroyed 1000 million birds.

By 1896 there were only 250,000 Passenger Pigeons left. They came together in one last great nesting flock in April of that year outside Bowling Green, Ohio, in the forest on Green River near Mammoth Cave. The telegraph lines notified the hunters and the railways brought them in from all parts. The result was devastating – 200,000 carcasses were taken, another 40,000 were mutilated and wasted. 100,000 newborn chicks not yet at the squab state and thus not worth taking were destroyed or abandoned to predators in their nests. Perhaps 5,000 birds escaped.

The entire kill of this hunt was to be shipped in boxcars to markets in the east, but there was a derailment on the line on the day of the shipping. The dead birds packed in the boxcars soon began to putrify under a hot sun. The diligent hunters' efforts were wasted: the rotting carcasses of all 200,000 birds were dumped into a deep ravine a few miles from the railway loading depot.

On 24 March 1900 in Pike County, Ohio, the last Passenger Pigeon seen in the wild was shot by a young boy. On 1 September 1914 in the Cincinnati Zoo, 'Martha' – a Passenger Pigeon born in captivity – died at 29 years of age. She was the last of her species. The impossible deed was done.

PIGEON HOLLANDAISE

Columbidae
Alectroenas nitidissima
EXTINCT *c.* 1826
Mauritius

The Dodo was not the only pigeon to suffer extinction on Mauritius. There was also the striking crested Pigeon Hollandaise, which is commonly called the Mauritius Blue Pigeon *(Alectroenas nitidissima).* Like the Common Dodo, it was endemic to Mauritius. This exotic Blue Pigeon did not suffer from the Dodo's inability to fly, nor did it nest on the ground. It was a graceful forest bird that fed on fruit, berries and seeds. It lived in large flocks and nested communally in trees. Again unlike the Dodo, this bird was delicious to eat. 'Shooting parties' were often organized by the resident Europeans for sport and food. There was also a peculiar catch-all bounty paid by the government from 1775 onward for the killing of all 'vermin'. 'Vermin' seems to have included pigeons, and practically every other wild creature on the island as well.

In the long run it was not the hunters, nor the imported dogs, cats, hogs and rats that finally exterminated the Mauritius Blue Pigeon. It was the pet monkeys which the colonists had released into forests and allowed to breed in vast numbers. This same predator has reduced that other endemic dove, the strange Mauritius Pink Pigeon, to less than 20 birds today. The monkeys, capable of searching out the pigeons' nests in the highest of trees, continually raided them for eggs. The result was that the Blue Pigeon could not sustain its numbers with the heavy mortality of both adults and young. The last specimen of the Mauritius Blue Pigeon was shot in 1826 in the Savanne Forest.

BONIN WOOD PIGEON

Columbidae
Columba versicolor
EXTINCT *c.* 1900
Bonin Islands, Northwest Pacific

In the June of 1827 Captain Beechey in the HMS *Blossom* sailed into the Bonin Islands, which lie about 800km (500 miles) south of Tokyo harbour. He landed on the main island of the central group and named it Peel Island. Captain Beechey was the first to make mention of what we now know as the Bonin Wood Pigeon *(Columba versicolor),* a distinctly different species from the related Japanese woodpigeon. It was generally a much paler bird than its relatives, but had a strikingly colourful back, which was described as 'metallic golden-purple'.

On the *Blossom*'s visit to Peel Island a considerable variety of birds was reported in great numbers, though shipwrecks had already resulted in numerous pigs and rats being released on the islands. Only 25 years later, in 1853 when Commodore Perry passed through the Bonin Islands en route to Japan, he stated that birds had become rare. At this time there were only 31 people on the Bonin Islands, but there were deer, goats, pigs, sheep and 'innumerable cats and dogs . . . where no mammals, except a bat had previously existed'.

In 1875 Japan annexed the islands. At that time there were only 100 people on the islands, but by 1900 more than 5000 inhabited them. This rapid population growth seems to have ended the Wood Pigeons' chances of survival. The forest fast disappeared. A man named A. P. Holst took the last specimen of the Bonin Island Wood Pigeon from the island called Nakondo Shima in 1889. Between 1900 and 1929 various collectors searched the islands in vain for the bird.

CHOISEUL CRESTED PIGEON
Columbidae
Microgoura meeki
EXTINCT *c.* 1910
Choiseul, Solomon Islands, West Pacific

Among the Solomon Islands of Melanesia is Choiseul, once home of the most spectacular of all the Solomon Islands birds. The Choiseul Crested Pigeon *(Microgoura meeki)* is known from only six specimens, gathered by the Australian, A. S. Meek, in 1904. At that time the natives were thought to be hostile and it seems Meek acquired the birds in trade from a village and consequently did not know exactly what locality they came from. Modern ornithologists surmise that this pigeon inhabited remote cloud forests in the island's interior.

The Choiseul Crested Pigeon was first named and described scientifically by Walter Rothschild, who bought some specimens in the year they were collected. Later expeditions in 1927 and 1929 by experienced collectors attached to the Whitney South Sea Expedition of the American Museum failed to find this beautiful pigeon.

Being one of the most exotic of the Solomon Islands birds, it was also one of the most specialized. It was largely ground-dwelling, and it fed and nested very close to the forest floor. At the time of the American Museum expedition, the Melanesian natives said that the bird hadn't been seen for many years, and they said themselves that introduced cats had killed them off.

TANNA DOVE
Columbidae
Gallicolumba ferruginea
EXTINCT *c.* 1800
Tanna, New Hebrides, West Pacific

South of the Solomon Islands, on the eastern rim of the Coral Sea, is the island of Tanna in the New Hebrides. Here lived another rare Melanesian pigeon, the Tanna Dove *(Gallicolumba ferruginea).*

Practically nothing is known about the history of this bird. All our information comes from the naturalist Johann Reinhold Forster and his son (the illustrator G. Forster) who were attached to the Second Pacific Expedition of Captain James Cook.

In his *Observations Made During a Voyage Around the World* of 1778, Forester wrote that on 5 August 1774 Cook's sloops, *Resolution* and *Adventure,* 'anchored in a port on the new island, which has a volcano. The natives call their island Tanna . . . its circuit appears to be about 24 leagues' [100km]. Forster spent a lot of time investigating the 'Doogoos', or hot springs near the tideline of the harbour and watching the volcano 13km (8 miles) away erupting and creating a cloud 'like a large cauliflower'. The expedition remained on the island for two weeks, during which time Forster discovered a dove unique to Tanna. On 17 August, Forster wrote: 'We came into a wood . . . where a dove of a new species was shot. It is a small dove, with both head and breast a rusty brown, but what marks it off from the other new doves is the dark green wings and the strangely yellow eyes.' Forster then follows this up with a much more comprehensive description noting in detail many other features, such as its black bill, red feet, grey belly and dark reddish-purple back. Later Forster made a study of the fruits and vegetables of the island. In the process he indirectly gives us a further detail concerning the bird's diet. He does this by telling us how he first discovered the wild nutmeg of Tanna 'without ever being able to find the tree'. It was found 'in the craw of the pigeon which we shot . . . which

Bonin Wood Pigeon
Columba versicolor
Extinct *c.* 1900
Choiseul Crested Pigeon
Microgoura meeki
Extinct *c.* 1910
Pigeon Hollandaise
Alectroenas nitidissima
Extinct *c.* 1826

disseminates the true nutmegs of the East Indian Isles.'

This is all we know of the Tanna Dove. It is only because of Forster's report and his son's illustrations that the world became aware of the bird's existence – and of its passing.

NORFOLK ISLAND PIGEON
Columbidae
Hemiphaga novaeseelandiae spadicea
EXTINCT *c.* 1801
Norfolk Island, New Zealand

Norfolk Island Pigeon
Hemiphaga novaeseelandiae spadicea

Accounts of the Tanna Dove and the Choiseul Crested Pigeon lead one to suspect that quite a number of rare island birds may have disappeared without even being noticed or noted by the men who invaded their territories.

We know, for instance, that on Napoleon's famous island of exile, St Helena, there was a blue dove which was apparently extinct by 1775. Yet so little is known of the species, that no name has yet been assigned to it.

Of other extinct birds, such as the Norfolk Island Pigeon *(Hemiphaga novaeseelandiae spadicea)* we know only marginally more. This subspecies of the New Zealand Pigeon inhabited the tiny Norfolk Island some 800km (500 miles) northwest of New Zealand. It is likely that cats, forest destruction and imported weasels, along with human hunting practices, were the causes of extinction. Further, it seems the Norfolk Island Pigeon could not adapt to imported plants as staples. The last recorded sighting was in 1801.

LORD HOWE ISLAND PIGEON
Columbidae
Columba vitiensis godmanae
EXTINCT *c.* 1853
Lord Howe Island, Tasman Sea

Another pigeon from an island in the Tasman Sea (between New Zealand and Australia) is now known to us only through a painting by Midshipman George Raper of HMS *Sirius*.

In 1790, the *Sirius* stopped at Lord Howe Island for water and supplies. It was here that Raper painted the Lord Howe Island Pigeon *(Columba vitiensis godmanae)*. It was not until 1915 that the painting was recognized for what it was, as no specimens or skins of this bird had ever been collected. Thus again, only by the sheerest chance was solid evidence of a bird's existence discovered.

The Lord Howe Pigeon was just one of an amazing variety of indigenous birds that lived on an island only 11km (7 miles) long and 1·5km (1 mile) wide. Early European observers never seemed to tire of their own amazement at the tameness of birds who had never before seen men. Yet their reactions to such unwariness were always predictably and brutally the same.

In the year 1788, the Surgeon Arthur Bowes of the *Lady Penrhyn,* on landing on Lord Howe Island, wrote: 'The sport we had in knocking down birds was very great . . . the Pidgeons were the largest I ever saw . . . When I was in the woods among the birds I could not help picturing to myself the Golden Age as described by Ovid – to see the Fowls or Coots some white, some blue and white, others all blue with large red bills . . . together with a curious brown bird about the size of the Land Reel in England walking totally fearless and unconcerned in all parts around us so that we had nothing more to do than stand still a minute or two and knock down as many as we pleased . . . if you missed them they would never run away . . . the Pidgeons also were as tame as those already described and would sit upon the branches of the trees till you might go and take them off with your hands . . . many hundreds of all sorts mentioned above together with Parrots and Parroquets, Magpies and other Birds were caught and carried on board our ship with the *Charlotte*.'

The captain of the *Charlotte,* Thomas Gilbert, wrote about the birds in his diary on the same day: 'Several of these I knocked down, and their legs being broken, I placed them near me as I sat under a tree. The pain they suffered caused them to make a doleful cry which brought five or six dozen of the same kind to them and by that means I was able to take nearly the whole of them.'

The toll of the European ships on the birds of the island was such that a Dr Foulis who lived on the island from 1844 to 1847, when the human population was still only 16, wrote: 'There are but few birds that belong to the island, the only valuable kind being a large blue pigeon . . .' It is no wonder that by 1853, at least, the Lord Howe Pigeon was no more. Surgeon Bowes' 'Golden Age' was over.

Traditionally the Dove is the sign of peace, innocence and gentleness; it is also symbolic of the presence of the Holy Spirit. According to medieval chronicles: 'the Dove is a simple fowl free from gall, and it asks for love with its eye.' For at least five thousand years Men have been keeping doves, and have loved them for their peaceful nature. Doves and pigeons feed from the hand of Man and have even adapted to 'cliff-dwelling' in the world's largest cities on man-made towers and buildings. Adaptation and tameness has frequently been their salvation.

It is strange, therefore, that these characteristics of fearlessness and naivety, when encountered by European men in the wild, have invariably been interpreted as stupidity. So we find that the characteristics which in one context are seen as praiseworthy, and in another damning, have resulted in the extinction of not less than one dozen species of 'gentle doves'.

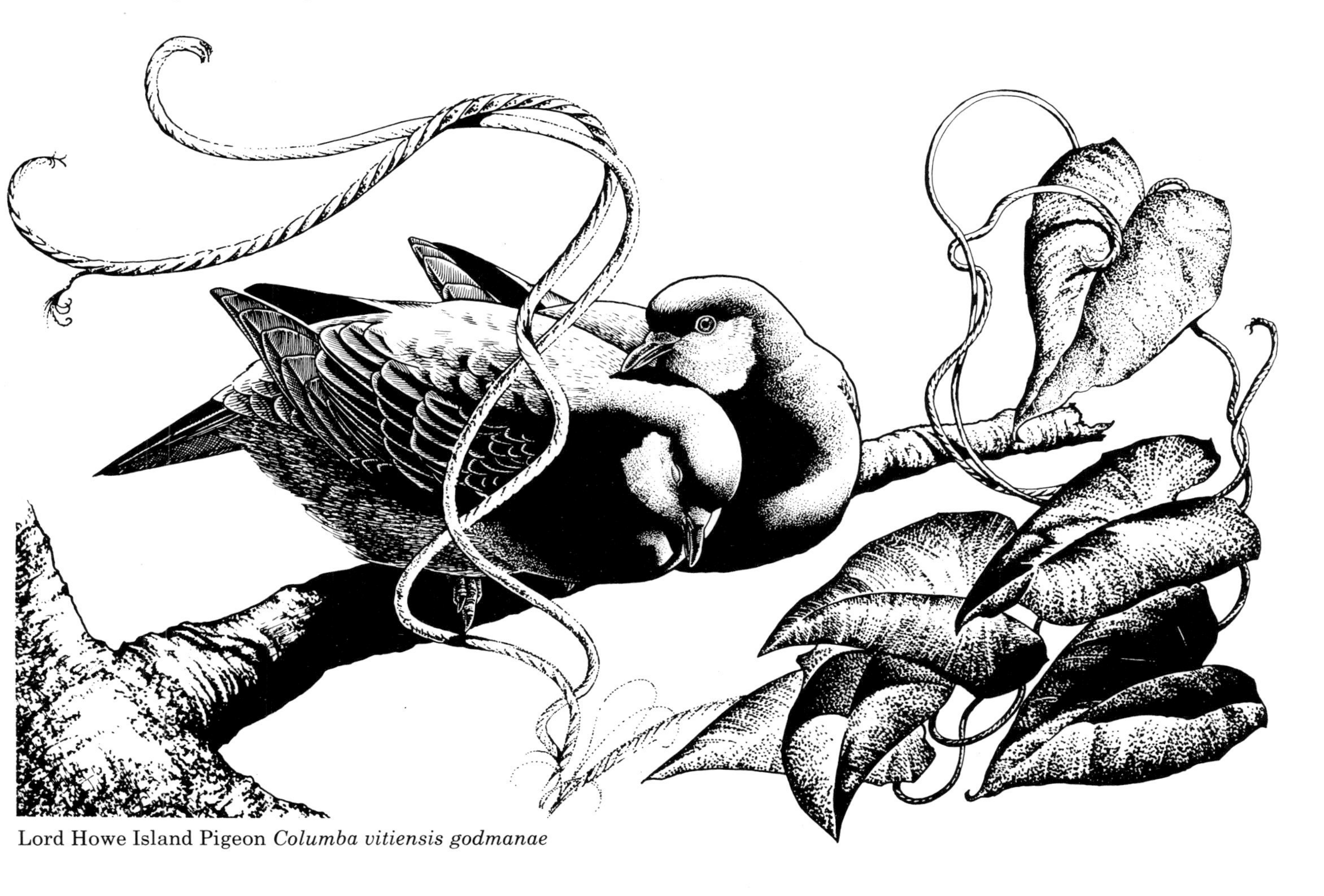

Lord Howe Island Pigeon *Columba vitiensis godmanae*

Great Auks *Alca (Pinguinus) impennis* Extinct c. 18

Birds of Sea, Shore and Marsh

THE SWIMMERS AND THE WADERS

It might have been someone's nightmare of hell. Men with faces and hands blackened by smoke worked with hooked poles and iron rakes over huge, boiling cauldrons. Blistering flames licked up into the air; the smoke was like pitch. Tens of thousands of large penguin-like birds stood in great pen enclosures where they had been herded. They stood erect like dwarfish men in a trance; they were packed so tight they could barely move.

They were white-breasted, wedge-beaked and black-backed. They were elegantly streamlined diving birds, but they were defenceless on land and they could not fly. The men who had herded them into the pens waded through the birds swinging spiked clubs. Killing or stunning a bird with each blow, they worked their way through the enclosures. Men followed the strikers, lifting the birds and throwing them over the enclosure walls into limp piles near the fires. There, another gang took the dead and wounded birds and either dropped them into the boiling cauldrons or threw them directly into the fire pits. There was no wood with which to feed the flames in this place, so the thick insulating layer of fat which protected the birds from arctic

GREAT AUK

Alcidae
Alca (Pinguinus) impennis
EXTINCT *c.* 1844
North Atlantic Islands

Great Auk
Alca (Pinguinus) impennis

waters served as fuel for the cauldrons where the feathers were separated from the birds' bodies. Men with rakes scooped the feathers from the surface and stuffed them in sacks. Men with hooked poles gaffed the naked birds and dragged them from the cauldrons down an embankment and onto the rocky sea shore where thousands of rotting corpses in piles awaited the next high tide that would wash them away.

It was a barren isle of rock less than one quarter mile wide and one half mile long. There was no plant growth on it because there was no soil, but millions of tons of dung, many feet thick from millenia of nesting birds, covered the whole island. During the nesting season, even while tens of thousands of birds were being slaughtered at one end of the island, hundreds of thousands were unwilling to abandon the eggs not smashed in the herding or taken by egg collectors. It seemed as though they sat in silent witness and awaited their turn.

It was not, of course, just someone's evil dream. This was a place called Funk Island, just off the coast of Newfoundland, in 1800. It was one of the nesting grounds of the bird called the Garefowl or, more commonly, the Great Auk *(Alca impennis)*. It was called Funk Island because of the overpowering stench ('funk') of the Great Auk 'industry'. Many contemporary witnesses described such slaughter-house scenes, and most have been recorded in S. Grieve's nineteenth century *The Great Auk, or Garefowl . . . its history, archaeology and remains*, which is still the best book on this bird. By 1844 there was not a single Great Auk left on Funk Island; nor was there one to be found anywhere in the world. And even today on this desolate isle the black pits of ashes and charred bone remain as a memorial to those industrious masters of the feather trade.

The Great Auk was the original Penguin ('white head' in Welsh), the penguins of the Southern Hemisphere having derived their name from European sailors who saw the similarity. It was the strongest and swiftest of northern swimming and diving birds, and almost immune from marine predators; while its rocky nesting sites in the rough North Atlantic were virtually inaccessible to men until historical times. It numbered in the tens of millions. From the tenth century, the birds were occasionally raided by sailors and fishermen as an alternative food source on some of their nesting sites in the Outer Hebrides and a few islets off Iceland, but their numbers were so great that this hunting was unlikely to have affected the populations to any extent.

A curious theory links the Great Auk with the discovery of North America. It proposes that the Icelandic Vikings, already familiar with the Auks and aware in a general sense of bird migrations, would have noticed the seasonal appearances and departures of vast rafts of the birds in their western waters. It would have been a small step for the Vikings to conclude that these Garefowl were travelling to other, similar lands in the west. Consequently they might have decided to follow the migrating legions of Auks, who would have led them first to Newfoundland and then further down the eastern coast of North America.

This is, of course, hypothetical. However, having now established that the Vikings were indeed resident in Newfoundland, we can be sure that they were the first Europeans to hunt the Great Auks in North America. The first actual records we have of the killing of

Auks off Newfoundland are from 1497 and the French ships which sailed to the astounding cod fisheries of the Grand Banks. They took the 'pingouins' in such huge numbers, that it was considered unnecessary to stock their ships with food for the duration of their stay off the Grand Banks: the ships could secure all the fresh meat and eggs needed by hunting these birds at their nesting sites. In 1534, Jacques Cartier visited an 'Island of Birds' off Newfoundland. His crew filled two boats with Auks in less than half an hour, and every ship salted down six barrelfuls. There were at least three nesting islands off Newfoundland at that time, as well as the Magdalens in the Gulf of St Lawrence, and other islands off the tip of Nova Scotia and in Massachusetts Bay.

Extensive though the killing of Great Auks for flesh and eggs was (a Captain Mood recorded taking 100,000 eggs in one day), it was not until the feather industry turned its attention to the birds that they began truly to head for extinction.

Around 1760, the supply of eider-down and feathers for feather beds and coverlets was exhausted, through overhunting of breeding ducks and the destruction of their nesting grounds along the whole of the east coast of North America. Consequently the feather merchants sent out crews of men to the Great Auk nesting grounds, and by 1810 Funk Island was the only West Atlantic 'rookery' left. The crews returned each spring until they had killed every bird.

In Europe the birds lasted somewhat longer off the Iceland coast, largely because their destruction was not so well organized, but still much the same unrestricted killing and theft of eggs resulted in a steadily declining population. And as their numbers grew less, the hunting of the birds increased, until by 1840 they were practically never seen – though ornithologists were by that time out to collect any that might remain. So it was that on 3 June 1844, on Eldey Island (often called Fire Island) off the coast of Iceland, three Icelandic fishermen discovered the last two living Great Auks. They were a breeding pair with a single egg. Jon Brandsson and Sigourer Isleffson killed the two adult birds with clubs. Ketil Ketilsson smashed the egg with his boot.

On 4 March 1971, the Director of Iceland's Natural History Museum paid £9,000 in Sotheby's auction rooms for a mounted specimen of a Great Auk. It was paid for by public subscription and was the highest price ever paid for a stuffed bird. The museum said it would have been willing to go to £20,000, if necessary, to acquire the specimen.

STELLER'S SPECTACLED CORMORANT

Phalacrocoracidae
Phalacrocorax perspicillatus
EXTINCT *c.* 1850
Bering Island and satellites, Bering Sea

When the Danish navigator Vitus Bering embarked in 1741 on his last voyage of discovery for Russia into the North Pacific, he was rapidly given a foretaste of the ill-luck that was to dog his expedition. In June his ship, the *St Peter,* lost contact for ever with its companion ship, *St Paul,* in wind and fog. Over four months later the *St Peter* drifted helplessly onto the rocks of an unknown island, its captain dying and its crew racked with scurvy and despair. Bering was buried there and both the island and the sea in which it lies have taken his name.

There followed for the survivors an eight month ordeal such as few men have known: dug into fox-holes in the sand they sat out the winter; short of driftwood for fires, dying of scurvy, harrassed night and day by the swarms of ravenous arctic foxes. A young naturalist, a previously aloof and arrogant man called Steller, was one of them. Now his true nature emerged: he saved lives, maintained morale – and kept his trained eyes open.

There was much for him to note in his journal, for Bering Island, never before visited, was rich with wildlife. Food, in fact, was no problem for the men: what they lacked was adequate shelter, warmth and, above all, vitamins.

Among Steller's discoveries of unknown animals were 'a special sea-eagle' identified later as *Haliaetus pelagicus;* 'a white sea-raven', a mystery bird, never again sighted; the astonishing and gigantic 'Sea Cow', *Hydrodamalis stelleri;* and the large cormorant, now called Steller's Spectacled Cormorant *(Phalacrocorax perspicillatus).*

In midwinter, when the sea otters and fur seals had become scarce and the flocks of eider ducks were far offshore, the crew turned to the Cormorants for food. Almost flightless, the ring-eyed birds were slow-moving and unsuspicious and they provided a substantial meal for little effort: 'They weighed 12–14 lbs so that one single bird was sufficient for three starving men.'

Though one might expect a fish-eater to be unpleasant tasting, Steller reported that the 'Katchadals' (Kamchatkans) in the crew cooked them *à la tziganne* (encased, feathers and all, in clay and baked in a sandpit) and that they were delicious and juicy.

Steller was the only naturalist to see the Spectacled Cormorants and they were known to the scientific world only through his description until 1837, when the first prepared specimens were brought in to Sitka, Alaska.

Meanwhile, however, Bering Island had become a regular stocking-point for whalers and sea-otter hunters. In 1826 the Russian-American Company imported Aleut natives to settle there and they esteemed the Cormorants a delicacy. Though Steller had described the birds as existing in 'most copious' numbers along the rocky shores, they were unable to adjust to these new pressures. Their last refuge was on one of their fox-free nesting skerries, Aij Kamen. By 1850 they were all gone.

The Spectacled Cormorant has the dubious honour of being one of the rarest museum specimens in the world: only six stuffed birds exist, and two imperfect skeletons, from the unsuspecting flocks which had watched the *St Peter* breaking up on the rocks of Bering Island in 1741.

ESKIMO CURLEW

Scolopacidae
Numenius borealis
EXTINCT *c.* 1970
North and South America

There could be no greater contrast with the flightless Cormorant of Bering island than the Eskimo Curlew *(Numenius borealis),* a little-known globetrotter whose passing was, in many ways, as dramatic as that of the Passenger Pigeon.

This little wader, perhaps 30cm (1ft) long with a 50mm (2in.) probing beak, pursued the most elaborate and hazardous migration cycle ever recorded. From its northern breeding grounds in the tundra of the Canadian sub-arctic, it followed a huge clockwise circle of flight through the western hemisphere: east through Labrador, down through the Atlantic and across the southern Caribbean, then flying on until it reached the Argentinian Campos south of Buenos Aires and possibly even further, beyond the Andes into Chile.

Steller's Spectacled Cormorant *Phalacrocorax perspicillatus* Extinct *c.* 1850

In the spring-time it completed the circle, crossing Texas and making its way through Kansas, Missouri, Iowa and Nebraska back to Canada. It seems to have travelled, with all its companions, in one gigantic flock.

The natural hazards of this cyclical journey were great enough. If there were storms in the North Atlantic (and in equinoctial September there often were) the Curlews could be blown off course either east or west. In the first instance they would come to land in New England and, until man arrived in numbers, that was endurable; but if they were blown west huge numbers would die, exhausted, in the ocean and the few recorded stragglers in Britain and on Atlantic ships were probably survivors of such disasters. At least, though, the northern journey overland was comparatively danger-free. Until, that is, the settlers arrived. And what did they see?

They nicknamed the birds 'prairie pigeons', recalling the sky-darkening flocks of Passengers they had seen in the east. In July 1833 Audubon had visited Labrador as the Curlews crossed the coast and made the same comparison. Prairie witnesses spoke of the great cloud of birds, flying in a loose wedge formation and 'wheeling together in any of their many beautiful evolutions'. But it is the *sound* of the Eskimo Curlew which comes down to us most vividly from the accounts of those who saw them. The Innuit name for the bird was *pi-pi-pi-uk,* in imitation of the call, and often the flocks could be heard before they came in sight.

'I saw a flock a mile long and nearly as broad', wrote a Dr Packard in 1860. 'The sum total of their distant notes resembled the wind whistling through a ship's rigging or at times the jingle of many sleigh bells.' Other observers spoke of 'a soft melodious whistle, *bee, bee*'; 'a low conversational chatter'; and 'a soft whistle as they landed and a chirping whistle as they walked about while feeding as if calling to each other'.

When they did land the sight can be imagined: a flock covering 40–50 *acres* was seen more than once in Nebraska. But what one must realize is that this may have been the total world population of the Curlews – all in one place and most vulnerable.

Until the great plains were settled this vulnerability only obtained in the pre-migration gathering grounds on the Labrador coast, and on the first leg of their journey in Newfoundland and Nova Scotia. But what they experienced there was a taste of things to come: 'The tide was rising and about to flood a muddy flat . . . where their favourite snails were. Although six or eight gunners were stationed on the spot and kept up a continual round of firing at the poor birds, they continued to fly distractedly above our heads, notwithstanding the numbers that every minute fell. They seemed in terror lest they should lose their accustomed fare of snails that day . . .'

When the Curlews reached Newfoundland they were stalked on their roosting grounds along the beaches just above the tide-line. Men would come among them in the darkness, dazzling them with lanterns and striking them down 'in enormous numbers' with sticks. The carcasses were salted down in barrels against the following winter.

When the 'doughbirds' (so called because they became immensely fat in their Labrador feeding) got blown in to New England by gales 'It was the signal for every gunner and market hunter to get to work . . . nearly all that remained on our shores were shot.' In the 1830s and 40s Eskimo Curlews came to Nantucket in such huge (and storm-exhausted) numbers that the island's supply of powder and shot ran out and the slaughter had to be 'interrupted'.

Until 1858 this East Coast abundance continued, but a sudden massive decline set in thereafter. Even in Labrador by the late 1870s the birds were a rarity and the last sizeable 'bag' of Curlews in New England was in the Fall of 1863. The last bird seen on the Halifax market in Nova Scotia was in 1897, and the last Labrador sighting was in 1912 when eight birds were seen at West Bay, of which seven were shot and their skins preserved.

Many puzzled over this disappearance. Some blamed Atlantic storms, others speculated that Eskimos had happened on the secret breeding grounds and slaughtered the lot! One would have thought that their own slaughtering offered explanation enough. But in fact they were right – the cause of the Curlews' disappearance did lie elsewhere, but not in the North. West, on the great plains and especially in Nebraska, the previously hazard-free northern route of the 'prairie pigeons' had become the stage for a squalid series of annual massacres, rivalling anything that colonists and cavalry were doing to the human and four-footed natives of the prairies.

'Hunters would drive from Omaha and fill wagonloads each, with sideboards on. Often they would be so well supplied with ammunition that they would dump piles of birds on the prairie to rot, as large as a couple of tons of coal, and then go back and shoot some more.'

Eskimo Curlew *Numenius borealis*

A professional class of market-hunters, too, who used the field glass, the new weapon of the Indian-killers, developed in Nebraska. They would spot flocks on the ground and approach within 23–27m (25–30yd) without difficulty. They would fire as the birds rose – one single shot was recorded to have dropped 28 Curlews stone dead, while wounded birds kept falling for the next half mile.

The Curlews would fly on two or three miles and the hunters would follow by horse and buggy. And so it would go on, the whole length of the state. Almost invariably the flocks were feeding on cutworms and the larvae and eggs of the Rocky Mountain Grasshopper: the greatest single enemy of the settlers who were turning the prairies into farmland.

In April 1900 Paul Hoagland was hunting with his father near Clarks, Nebraska. They put up about 70 birds near the road and followed them to a newly-ploughed field where they shot 34 with four shots. Eleven years later the same man saw eight birds and killed seven of them. He saved two of them for stuffing. The Eskimo Curlew was never seen alive again in Nebraska.

Since 1900, perhaps 20 specimens of the Curlew have been collected by ornithologists. In the late 1950s and early 1960s there were a number of sightings (and even one photograph) of single migrating birds in Texas, and in 1964 a Curlew was shot in Barbados in the West Indies. However, the vast flocks of the Eskimo Curlew are gone forever. The species has been virtually extinct since the beginning of the century. There was no hope for these desperate survivors whose instinct to migrate the length of two continents, without the protection that large numbers affords, doomed them to follow their ancestors.

BONIN NIGHT HERON

Ardeidae
Nycticorax caledonicus crassirostris
EXTINCT *c.* 1879
Bonin Islands, Northwest Pacific

The Bonin Islands seem to have been a complete sanctuary from man until the wrecking of the whaler *William* in 1826 and the brief establishment of a colony on Peel Island by its survivors. The shoals of voracious sharks encountered in its shallow waters may partly explain this isolation. As far as we know only two boats had been there before the *William* (a Japanese junk wrecked in 1675 and an English supply ship calling for water in 1825). But 40 years after the castaways of the *William* came ashore all four native bird species had disappeared from Peel.

Among these was the 'handsome brown heron with white crest' noted by Captain Beechey of the British ship, the *Blossom*. This spectacular 'night heron' was common along the rocks and in the marshes, and made its nest in low trees near the shore. The Bonin Night Heron *(Nycticorax caledonicus crassirostris)* fed on fish, insects and possibly young turtles. It stood 60cm (2ft) tall and two long white plumes reached from its black-crowned head to its back. The back was cinnamon brown and the belly white. Set off by orange legs and feet and with its great black spear-beak, it was a gorgeous creature indeed and obviously fascinated the *Blossom's* crew. Mr Collier of that ship wrote: 'Three *(sic)* white tapering feathers formed a crested plume on these birds but many were shot without the crest. They frequented the rocks on the seashore. They had one Caecum'.

Yet it's unlikely that avid collectors and amateur anatomists were responsible for the Bonin Night Heron's extinction. Not

Bonin Night Heron *Nycticorax caledonicus crassirostris* Extinct *c.* 1879

directly, that is. Cats and rats, of the numerous feral introductions, are more likely to have got at the eggs; and the nesting sites were quickly cut down by the settlers.

This Heron was marked off distinctively from related subspecies in the Celebes, New Caledonia and the Phillipines, by its massive and straight bill, as noted by Mr Vigors of the *Blossom,* the first naturalist ever to see it. This feature explains the Latin name, *crassirostris*.

In 1879 the last specimen of the Bonin Night Heron was taken from its final refuge, the northern island of Nakondo Shima. All searches for it since have been fruitless.

GUADALUPE STORM PETREL

Hydrobatidae
Oceanodroma macrodactyla
EXTINCT *c.* 1911
Guadalupe, Mexican Pacific

Many point to the Great Auk and the Spectacled Cormorant and claim that, like the Dodo, these birds did not survive because of the 'evolutionary mistake' of foregoing their ability to fly. If that is so, what is the 'lesson' of the Guadalupe Storm Petrel *(Oceanodroma macrodactyla)*? This diminutive pelagic bird enjoyed the freedom of three elements: land, sea and air. Furthermore, like other species of Storm Petrel, these birds were protected from human interference by the widespread superstition that 'Mother Carey's Chickens' (as they are often called) are the embodiment of the souls of drowned mariners. The Guadalupe Storm Petrel's only 'weakness' was its choice of nesting site. It might otherwise be a common sight still off the coast of Baja California.

Guadalupe Storm Petrel
Oceanodroma macrodactyla

The Storm Petrels are familiar to sailors the world over, fluttering like dark butterflies around plankton-slicks far out to sea, or actually 'paddling' on the surface of the ocean with their small webbed feet. Like the shearwaters they nest in burrows on offshore islands, and it was this species' misfortune to have as its base the island of Guadalupe, 290km (180 miles) southwest of San Diego, in the Pacific. If for nothing else, this island would be notorious for the effect of introduced cats on its avifauna: 40 per cent of its birds were lost to this predator alone.

The Guadalupe Storm Petrel was one of the victims. The nesting burrows were at the north end of the great central ridge which characterizes the island, just where the land falls from 1400m (4500ft) to about 750m (2500ft), and where a grove of white pines supersedes the endemic cypresses and sage brush. The tunnels were lined roughly with leaves and pine-needles, and a single egg was laid, marked with a striking wreath of lavender-coloured stipples at its larger end. The adults would feed at sea, leaving the chick unattended during the day, and would fly back to the island at night, their crops full of predigested food. The petrel colony on Guadalupe was probably never more than a few hundred strong, and there is no evidence of decline before the late nineteenth century.

But the Storm Petrel has not been sighted since 1911. Cats were seen hunting along the ridge in the 1890s and dozens of the birds, both adult and unfledged, were found scattered and torn to pieces around the burrows.

The little carcasses were mute testimony to the defencelessness ashore of the 'chickens' who were at home in the wildest offshore weather, and who treated the giant swells of the wide Pacific as their playground.

Captain James Cook's great voyages through the 'South Seas' added enormously to the sum of human knowledge about the world. Cook carried with him zoologists, botanists, linguists and meteorologists: thorough and literate men who were, by and large, humane too.

But this opening up of the world had its immediate price. Sailors brought disease, or transferred it from one island to another, and – often inadvertently – the human and animal populations of the islands were nudged, one way or another, towards destruction.

Rats came off any European ships which could find inshore anchorage and the damage they have done is witnessed to repeatedly. But Cook's ships also carried pigs, goats and cattle, with the aim of benefitting native peoples and future European sailors.

We cannot guess whether the rats or the livestock were chiefly responsible for killing off the little white-winged Sandpiper of Tahiti, *Prosobonia leucoptera*. Certainly it was not over-zealous human collectors, for only three specimens were taken by Forster and then by Anderson on the two voyages.

This shy wader frequented the banks of small streams. When approached it would run off quietly through grasses or bush and then wing its way swiftly back towards the shoreline, whistling shrilly as it escaped. From this call the Tahitians named it 'torote' and the Mooreans 'te te'. It was, as a ground-nester, vulnerable both to egg-predation and to habitat destruction; and the banks of freshwater sources are often the first places to be colonized by animals and men.

In any case, the bird was never reported again after Cook's last voyage, and the apparently identical Moorean (or Eimean) species is only known to us through a painting in the British Museum.

TAHITIAN SANDPIPERS

Scolopacidae
Prosobonia leucoptera
EXTINCT *c.*1800
Tahiti
Prosobonia ellisi ?
EXTINCT *c.* 1800
Moorea, Society Islands

Tahitian Sandpiper *Prosobonia leucoptera*

Quelili *Polyborus lutosus* Extinct *c.* 1900

The Fierce and The Wise

THE BIRDS OF PREY

Among the most spectacular of all birds are the raptors, the so-called birds of prey: hawks, vultures, eagles, condors and owls. Today nearly 30 species of these birds are numbered among the most critically threatened creatures in the world. Apologists for the extermination of animals, as shown in the Doves chapter, have often given passivity and stupidity as reasons for the ease with which Man has disposed of them. These terms, highly debatable when applied to any animal species, are completely inappropriate when applied to the birds of prey. Hawks have become metaphors for ferocity and owls for wisdom, yet these admired characteristics have done little to save at least a dozen raptors from extinction.

The two largest flying birds on earth, the majestic Andean and California Condors, with wingspans of 3·4m (11ft), are both endangered. The great black and white Andean Condor, which for a thousand years has been venerated by the native Peruvians, is now hunted for its feathers (which are made into tourist junk) or stuffed by taxidermists who supply these 'trophies' to wealthy collectors. The California Condor's plight has been critical for at least a

century. This gentle giant was strictly a carrion feeder that never killed its own prey. It had no natural enemies except Man who, during the nineteenth century, ruthlessly hunted it. Since 1900 it has been on the razor-edge of extinction – only conservation measures and its longevity (up to 50 years) have allowed its numbers to remain at about 50 birds over these last 80 years.

Little hope as there may be for the Condors, there is even less for the majestic Monkey-eating Eagle of the Philippines. There are perhaps 100 pairs of these eagles left, and it is almost certain they will become extinct within this century, largely because of the fashion for displaying them as mounted trophies in wealthy Philippino homes.

Among the rarest birds in existence is the Mauritius Kestrel: by 1976 there were exactly six living birds in the world. Other endangered raptors include the once common Southern Bald Eagle, the Imperial Eagles, the Black Vulture, the Lammergeier, the European Griffon and numerous other eagles, vultures, hawks, falcons and owls. All of these birds are of striking individuality and beauty.

QUELILI

Falconidae
Polyborus lutosus
EXTINCT *c.* 1900
Guadalupe, Mexican Pacific

The magnificent large brown hawk called the Quelili or Guadalupe Caracara *(Polyborus lutosus)* was one bird of prey that did not survive human persecution. It was deliberately and knowingly extirpated by man.

The Quelili was a distinctly different bird from other Caracaras in North and South America, and is thought to have been almost identical to the original Pleistocene ancestor of these modern birds. Though classified among the falcons, the Quelili's similarities are technical and not immediately apparent. This was no stooping, high-speed hunter, like the peregrine or the lanner. The English-speaking Guadalupe islanders called it the 'eagle' and, with its broad wings and extremely sociable behaviour, it rather resembled a small, elegant vulture.

The Quelili fed on carrion, small birds, mice, shellfish, worms, insects and supposedly – a dubious belief which proved its death warrant – on young goats. The significance of this can only be explained in the context of the dismal and cautionary history of the island of Guadalupe.

Often confused with the Guadeloupe in the West Indies, Guadalupe lies 225km (140 miles) off the coast of Baja California, 290km (180 miles) southwest of San Diego. About 32km (20 miles) long by 9·5km (6 miles) wide, it has a great central volcanic ridge, rising to 1535m (5000ft) and running north-south down its entire length. Towards the ridge's north end is, or was, a large grove of endemic cypress trees, while white sagebrush and pines flourished on the weathered lava flows of its steep sides.

The original vegetation of Guadalupe has almost all gone, victim of goats introduced to the island in the early eighteenth century. There are now, incredible as it may seem, 50,000 goats on the island. The herded goats support the farmers, the feral ones hunters, and the island is, consequently, very little more than a close-cropped goat walk. Introduced cats are also numerous and, together with the goats, have been responsible for the disappearance of half the island's breeding birds – the endemic Guadalupe Storm Petrel of the previous chapter among them.

It was the goat-herds who waged unremitting war on the Quelilis. If they were right in their claim that the 'eagles' took healthy kids, the toll must have been small indeed in relation to the huge numbers of goats. Nevertheless, the Quelilis were hunted mercilessly until there were only a handful of the birds left. By the 1860s poison and shooting had brought them to the brink of extinction and at this point the interest of American collectors (however scientifically well-meaning) tipped the balance.

In 1875 the ornithologist Palmer, knowing the Quelili's rarity, went over and collected a dozen skins. He also left us one of the few detailed descriptions of the living birds: 'In fighting among themselves and when excited they make a curious gabbling noise . . . with an odd motion of the head, the neck being first stretched out to its full length and then bent backwards till the head almost rests upon the back. The same odd motions and noises are made when they are about to make an attack upon a kid. When surprised or wounded, they utter a loud harsh scream something like a bald eagle.'

The collectors' interest had its effect. The last chapter in the story of the Quelili's extinction is a blend of squalor and mishap. In 1897 a goat-hunter captured four 'eagles' and brought them over to the mainland on the schooner *Francine*. He set them up on public display in a large cage in San Diego, determined on making the greatest possible financial profit from their sale. The exotic birds attracted a lot of attention and their captor was interviewed by a city newspaper. He was Harry Drent, a drifter round the world and at that time a hunter of the goats which had gone wild along Guadalupe's ridge-top. This was his story: 'It's a trick I learned in South Africa. The first bird I winged with a shotgun. I then made him a prisoner and staked him near a large boulder. I then took a string, fastened it to a stick and made a loop similar to a cowboy's lariat. I then hid myself behind the rock, knowing the other birds would come to the captive. I threw the rope and captured a second bird. I then made him a prisoner with the other and . . . secured four out of the seven remaining birds on the island . . . These were the first ever taken alive. I have been offered $100 for the four but won't sell them.'

The $100 offer may well have come from the Smithsonian Institute, which was alerted and interested, but Drent was holding out for $150 *per bird* – an extremely large sum in 1897. Within a month the hawks had all died and what Drent got for the carcasses is not recorded.

His comment on the number of Quelilis remaining on Guadalupe is revealing, though obviously mistaken. But his venture, or perhaps his misleading bragging on his return to the island, seems to have alerted other islanders to the possibility of making Yankee dollars from their traditional enemy. Later that same year a fisherman brought another live 'eagle' over to San Diego on his boat and demanded the same price that Drent had – $150.

Piqued by the discovery that such lordly rewards were not forthcoming, this man cut the Caracara's wings off and threw the body into the sea. Some young boys found it bobbing in the water near the bay shore, retrieved it and took it to Frank Holzner, an eminent taxidermist whose 'studio' was well known in the city. Holzner, in fact, had got hold of at least two of Drent's Quelilis.

The taxidermist tracked down the fisherman, just as he was leaving for Guadalupe, and persuaded him to part with the wings. He performed what was, by all accounts, a miracle of reconstruction or salvage: sewed the wings back on the body and put 'The Last Guadalupe Eagle' up on display in his emporium. Two weeks later the store burnt down and everything in it was destroyed.

But in fact there were still Quelilis on the island – one flock of them – and the story ends with the rueful account of R. H. Beck, a qualified ornithologist and avid collector who was not, it seems, *au fait* with recent literature on the rare hawk: 'On the afternoon of 1 December 1900, just after landing on Guadalupe, I saw a flock of [Quelilis] approaching. Of eleven birds that flew towards me nine were secured. I only saw those eleven, but judging by their tameness and the short time I was on the island, I assumed at the time that they were abundant.'

The collector's euphemism 'secured' here means shot and killed. We cannot guess whether the two remaining Quelilis were wounded or not – but they and their kind were never seen again.

PAINTED VULTURE

Cathartidae
Sarcorhamphus sacra
EXTINCT *c.* 1800
Florida, USA

The history of the colourful Painted Vulture *(Sarcorhamphus sacra)* of Florida is far more elusive than that of the Guadalupe Quelili. Indeed it is a bird whose very existence has been hotly debated.

In the *Travels* of the naturalist Bartram, published in 1791, there is a description of a voyage along the St John's River above Lake George in Florida. During that journey he notes the habits of the Black Vulture and the Painted Vulture: '. . . a beautiful bird, near the size of a turkey buzzard, but his wings are much shorter, and consequently he falls greatly below that admirable bird in sail. I shall call this bird the painted vulture. The bill is long and straight almost to the point, when it is hooked or bent suddenly down and sharp; the head and neck bare of feathers nearly down to the stomach, when the feathers begin to cover the skin, and soon becomes long and of a soft texture, forming a ruf or tippet, in which the bird by contracting his neck can hide that as well as his head; the bare skin on the neck appears loose and wrinkled, which is of a deep bright yellow colour, intermixed with coral red; the hinder part of the neck is nearly covered with short, stiff hair; and the skin of this part of the neck is of a dun-purple colour, gradually becoming red as it approaches the yellow of the sides and forepart. The crown of the head is red; there are lobed lappets of redish orange colour, which lay on the base of the upper mandible. But what is singular, a large portion of the stomach hangs down on the breast of the bird, in the likeness of a sack or wallet and seems to be a duplicature of the craw, which is naked and of a redish flesh colour, unless when it is loaded with food, (which is commonly, I believe, roasted reptiles) and then it appears prominent. The plumage of the bird is generally white or cream colour, except the quill feathers of the wings and two or three rows of the coverts, which are of a beautiful dark brown; the tail which is large and white is tipped with this dark brown or black; the legs and feet of a clear white; the eye is encircled with a gold coloured iris; the pupil black.'

Painted Vulture *Sarcorhamphus sacra* Extinct *c.* 1800

This great white vulture was never conclusively reported again and after his death Bartram was attacked as a liar who described either a mythical bird or one constructed of other birds' parts. In fact, Bartram was a thorough naturalist and a man of unimpeachable honesty. Moreover, this is an extremely detailed and accurate description of the King Vulture *(Sarcoramphus papa)* that ranges from central Mexico to Argentina. However, the apparently smaller size of Bartram's Painted Vulture, the longer bill and the white tail and white feet (the King Vulture's tail and feet are black) would suggest Bartram was correct in assigning it a distinct specific name.

In 1936 Francis Harper discovered Bartram's original on-the-spot journal notes, and became convinced that the vulture description was valid and accurate. The journal read: 'The Croped Vulture. This is a very beautiful bird, not quite so large as the Turkey buzzard, they are chiefly white, the back and wings of a deep nut brown, the bill yellow, legs white, the head and part of the neck bare of feathers covered with a naked skin of vermilion colour, what is remarkable in the Bird their craw or stomach hangs like a pouch or purse bearing outside on the breast and bare of feathers. When the vast meadows and Savannahs of Florida are set on fire, they gather in flocks to the new burnt ground where they feed on the roasted snake frogs lizards, turapins and other reptiles, where I had an opportunity of getting one.' This entry dates from the spring of 1775 and two things in particular convinced Harper: one, the fact that Bartram actually had a specimen to hand and, two, the 'stomach' hanging down was evidently meant to mean the crop – which does bulge and remain stretched in an adult King Vulture. Also he found that Bartram had not personally supervised the text of his published travels.

Descriptions of the bird by most other writers seem to derive from Bartram's book rather than from observation, but we do know that Du Pratz, in his *Histoire de la Louisiane* of 1758, was not drawing on the Englishman's publications. Du Pratz calls the Vulture 'king of birds' and says it is: '. . . smaller than the eagle of the Alps; but it is a much finer bird, being almost entirely white, and having only the extremity of its quills black. As it is rather rare, it is prized among the natives, who pay a high price for the wing quills as an adornment of the "peace-pipe".'

Writers since Francis Harper remain divided on certain points of classification for this bird. Many believe that it was a relic race of King Vulture isolated in Florida – and thus conclude that Bartram was wrong about the white tail and white feet. (Surely an unlikely lapse of concentration in such a detailed first hand description.) Others, more plausibly, believe the Painted Vulture was either a diverging sub-species (*S. papa sacra*) or a full species (*S. sacra*). Whatever the case, we can be certain that there was a smaller, white-tailed relative of the King Vulture in Florida – and possibly, Bartram implies, in Carolina too – in some numbers till the latter part of the eighteenth century at least, when severe frosts may have added to its problems. A report from the Schonbrunn Zoo in 1913 found that King Vultures suffered more from frost than any old-world vultures did, and one of their specimens had lost all its toes. It is likely, then, that the Painted Vulture's disappearance was inevitable and only hastened somewhat by Man's encroachment on its habitat.

Another extinct and mysterious bird of prey was the Madagascar Serpent Eagle (*Eutriorchis astur*), which inhabited the dense, humid rainforests of northeast Madagascar and was probably never common. It was the only species known of its genus.

The Franco-Anglo-American Zoological Expedition of 1929 collected two specimens from the vicinity of Maroantsetra, one near sea level and the other at about 600m (1970ft) altitude. The first was shot as it flew up a trail through riverside second-growth trees, the other as it perched on a high limb in dense forest on a hill-ridge.

It was a medium-sized hawk with short, rounded wings and a very long, rounded tail. Its grey plumage was heavily barred throughout, but the bars were less noticeable on its dark back. It had a cowled head, yellow eyes and, of extant birds, most resembled Henst's Goshawk (*Accipiter hensii* – also of Madagascar), although it was larger, paler and longer-tailed.

It is presumed the Madagascar Serpent Eagle shared the habits of other African Serpent Eagles, and hunted snakes, lizards and frogs from high perches; these birds being particularly noted for their ability to hunt and kill large venomous snakes. (The stomach of one of the Maroantsetra birds contained part of a very large chameleon.) Nearly all kill their own food and refuse carrion. To achieve this the Serpent Eagles are equipped with acute vision, and particularly strong and heavily-scaled legs and toes, as well as the powerful beak and claws of all the hawks.

Although very little has been learned from observation of this species, there seems to be small doubt of the reason for its extinction. Its only habitat, the humid rainforests to 550m (1800ft) altitude, have been largely cut down and destroyed, and with the dense forests the bird has also vanished.

MADAGASCAR SERPENT EAGLE

Accipitridae
Eutriorchis astur
EXTINCT *c.* 1950
Northeast Madagascar

Madagascar Serpent Eagle
Eutriorchis astur

For the Forest Spotted Owlet or Blewitt's Owl (*Athene blewitti*) of India, we have practically no records of observed behaviour. This Owlet inhabited the dense jungle areas of central India, near Sambalpur and Marial. It has not been seen since 1914, when the German collector Minertzhagen shot one in the now vanished Mandvi forest, 320km (200 miles) north of Bombay.

The Forest Spotted Owlet was a small owl about 23cm (9in.) long with rounded wings. It was to be found in moist, deciduous jungle and among groves of wild mango, and seemed to favour the neighbourhood of streams. It was apparently not nocturnal: 'some specimens were shot in heavy jungle below the Satpura Hills late in the morning while sitting alone on the exposed tops of thin trees.' In fact, though, less than a dozen specimens have ever been taken, all exhibiting one physical peculiarity: the quills on the heavily feathered feet present a jagged appearance where the filaments of the feathers have been somehow worn away.

In February 1975 the foremost authorities on Indian birds, Dillon Ripley and Salim Ali, set out to find the owl if it still existed. They found that the Mandvi jungle had been completely cleared, so they searched carefully through the forested areas of the Mahamadi River. They carried equipment to play tape-recordings of owl cries 'at dusk and in the evening' and called forth every known owl except the one they were hoping to find.

FOREST SPOTTED OWLET

Strigidae
Athene blewitti
EXTINCT *c.* 1914
Central India

Forest Spotted Owlet
Athene blewitti

LAUGHING OWLS

Strigidae
Sceloglaux albifacies albifacies
EXTINCT *c.* 1900
South Island, New Zealand
Sceloglaux albifacies rubifacies
EXTINCT *c.* 1900
North Island, New Zealand

In New Zealand there lived a peculiar endemic bird called the Whekau or Laughing Owl. With a total length of 40.5cm (16 in.), the Whekau was the largest of New Zealand's owls. There were two closely related but distinct forms: the South Island Whekau (*Sceloglaux albifacies albifacies*), and the North Island Whekau (*S. a. rubifacies*). Both have been extinct since the turn of the century and the complex forces which caused their decline were triggered by the settlement of New Zealand by Europeans and their attendant pets, pests and clearances.

The physical characteristics of these owls go some way towards explaining their vulnerability. They were more hawk-like than other owls, with prominent beaks, small heads and swollen nostrils, but at the same time they had very short and feeble wings for which they compensated with long legs and short toes. They were, in other words, highly specialized for ground-hunting. But this, as W. L. Buller, the first ornithologist to discuss them, pointed out, was an anomaly: 'strange that a bird formed specially by nature for preying on small quadrupeds should exist in a country that does not possess any'. (The Kiore or Polynesian Rat was introduced by the Maoris.)

Almost everything we know about the Whekau is due to one enthusiast, W. W. Smith of the Albury estate near Timaru, who wrote extensively about his experiments in capturing and breeding the 'laughing owls'. He traced the owls to their nesting crevices in limestone cliffs and burned dry tussock grass at their mouths till the little caves filled with smoke: 'After trying a few crannies, I found the hiding place of one, and after starting the grass, I soon heard him sniffing. I withdrew the burning grass, and when the smoke had partly cleared away, he walked quietly out, and I secured him. I obtained four birds by this means . . .'

Smith built a house for his owls and found that they became tame very easily, though the females hid at the slightest noise, unlike the larger and bolder males. They bred readily and during hatching the male supplied food attentively to his mate, as well as taking occasional turns at brooding. The owls ate beetles, lizards, mice, rabbits and mutton supplied by Smith, but preferred young and half-grown rats above all. He found them slow and clumsy in capturing live prey, but blamed this on their lack of exercise and freedom.

Every evening, upon waking, the owls joined in a 'peculiar hailing call' described as being exactly like two men calling 'coo-ee' to each other in the bush; and often these calls attracted wild owls who would come flying high over the owl house 'laughing' to the captives.

This 'laughing' sound has been variously described: one witness wrote of the 'loud cry made up of a series of dismal shrieks frequently repeated' and apparently this outburst came only when the birds were on the wing and generally on dark or drizzly nights, or immediately preceding rain. In a more homely vein, though, Smith describes the multitude of sounds made by his Whekaus: '(they) chuckle like a turkey, mew like a cat, yelp like a puppy and whistle tunelessly.'

Smith's studies of the owls took place in the early 1880s, but he eventually moved to the city and what became of the birds is not known. Thanks to him we do have a vivid picture of these endearing little predators: 'They have been', he said, 'a great source of pleasure and instruction to me.'

South Island Whekau *Sceloglaux albifacies albifacies* Extinct *c.* 1900?

COMORO SCOPS OWL

Strigidae
Otus rutilus capnodes
EXTINCT *c.* 1890
Anjouan, Comoro Islands, Indian Ocean

Unfortunately, few of the smaller extinct owls have had such observant chroniclers of their lives and habits as the Whekau did. On the island of Anjouan, in the Comoro Group off the east coast of Africa, the Comoro Scops Owl (*Otus rutilus capnodes*) passed into extinction without notice. Although related owls are to be found on neighbouring islands, the Comoro Scops Owl is known only from specimens collected by a man named Humblot in the 1880s. It is possible that this collector exterminate the rare bird, or at least over-collected almost to that point. There are two specimens in New York and 29 in Paris and London, all taken by Humblot, and we know that the great majority of his specimens were dispersed throughout various French provincial museums.

BURROWING OWLS

Strigidae
Speotyto cunicularia amaura
EXTINCT *c.* 1900
Antigua, Nevis and St. Kitts, West Indies
Speotyto cunicularia guadeloupensis
EXTINCT *c.* 1900
Marie Galante, West Indies

Guadeloupe Burrowing Owl
Speotyto cunicularia guadeloupensis

In the West Indies by 1900, two subspecies of the Burrowing Owl had become extinct. These were the Antigua Burrowing Owl (*Speotyto cunicularia amaura*) and the Guadeloupe Burrowing Owl (*S. c. guadeloupensis*). Both were victims of the introduced mongoose.

The effect, often devastating, of mongoose introductions to Caribbean islands must be seen in a certain geographical context: the eminent ornithologist J. C. Greenway proposes a rule for nearly all these islands, 'that birds disappear where humans are numerous and forests relatively small'. Thus he points out that Hispaniola, with 14.3ha (5.8 acres) of forest per human being, has not yet lost a single bird species though it has a large population of mongooses and rats.

It is, then, in the smaller or flatter islands where forests are quickly destroyed by settlement and cultivation, that the introduced pests have the advantage over native, ground-nesting birds. The Guadeloupe form of the Burrowing Owl was actually restricted to the small satellite island of Marie Galante; while the Antigua form was found on Antigua, Nevis and St Kitts. All four islands conform to Greenway's rule: they have high population density and considerable forest destruction.

Both owls disappeared about the same time, at the end of the nineteenth century, almost immediately after mongooses were introduced. They were probably never numerous birds: only six specimens exist of the Guadeloupe owl, all taken early in the century by L'Hermenier, and only five of the Antiguan birds. Both were significantly darker in colouring than other burrowing owls and the Antiguan owl was unusually small.

Nesting as they did in burrows dug out of banksides, the owls and their clutches of two to five eggs were vulnerable to both rats and mongooses. Long-legged birds, they were agile on the ground and active both by day and night. They had the habit when disturbed of staring fixedly at the intruder and bobbing up and down, rather than escaping. When they did take to the wing their flight was rapid and 'bounding' but was rarely sustained for more than a few yards

They uttered a variety of chattering sounds but their most characteristic cooing calls earned them the nickname among French-speaking Guadeloupeans of 'Coucou terre'.

The numerous bird casualties of the Mascarene Islands included at least three owls: Commerson's Scops Owl (*Scops commersoni*), the

Rodriguez Little Owl (*Athene murivora*) and the Mauritian Barn Owl (*Tyto sauzieri*). None is known from specimens and we have to rely on descriptions, drawings and the evidence of subfossil bones.

COMMERSON'S SCOPS OWL
Strigidae
Scops commersoni
EXTINCT *c.* 1850
Mauritius

Commerson's Owl from Mauritius, was unusually large for a Scops, being nearly two feet long. Some writers suggest that it was a subspecies of the Madagascan Long-eared Owl, but despite its prominent ear tufts, its unfeathered legs call this relationship in doubt.

The Mauritian-born naturalist, Julien Desjardins, supplies us with the only real evidence for its historical existence, aside from one inept drawing by a seaman named Jossigny. Desjardins wrote: 'In September 1837 many residents of La Savane told me they had seen owls in their forests; Dr Dobson, of the 99th Regiment, assured me he had shot one in the wood of Curipipe.'

RODRIGUEZ LITTLE OWL
Strigidae
Athene murivora
EXTINCT *c.* 1850
Rodriguez

The Rodriguez Little Owl is known only from leg and beak fragments. The anonymous *Relation de l'Île de Rodrigue* describes it as 'very like the brown owl, and eats small birds and lizards. Mostly they dwell in trees, and always utter the same cry at night time when they expect fine weather. However, when they think the weather will be poor you do not hear them.'

MAURITIAN BARN OWL
Tytonidae
Tyto sauzieri
EXTINCT *c.* 1700
Mauritius

The Mauritian Barn Owl is known from bones found in the Port Louis area. Some authors believe there were two species on Mauritius, giving the second one the name Newton's Barn Owl (*Tyto newtoni*). In any case, no barn owls survived there after 1700. Like all other birds of prey on the Mascarenes, they would have been given the catch-all nickname, *mangeurs des poules*, and persecuted accordingly.

Rodriguez Little Owl *Athene murivora*

Commerson's Scops Owl
Scops commersoni

Carolina Parakeets *Conuropsis carolinensis carolinensis* Extinct *c.* 1914

Talking Birds

THE PARROTS AND THEIR ALLIES

On the first of September 1914, the last Passenger Pigeon was found dead in its cage in the Cincinnati Zoological Gardens. Before the month was out, in that same institution the last Carolina Parakeet fell from its perch in a nearby enclosure and died.

The Carolina Parakeet (*Conuropsis carolinensis*) was a relatively small parrot – about 30cm (12in.) long and 280gm (10oz) in weight. Its body was green and yellow, its tail long and pointed, and it had an orange-yellow head. It was the only parrot native to the United States but until the later part of the nineteenth century it was extremely common in the eastern deciduous forest, especially in densely-wooded river-bottoms.

The Parakeets' decline almost exactly matched that of the Passenger Pigeons. Similarly the cause of their extirpation was almost exclusively overhunting by men who slaughtered them for sport, food or feathers. During the nineteenth century they were commonly kept as cage birds, and so trappers took many thousands to sell as pets, even though they hardly ever bred in captivity. Farmers, as well, killed large numbers. The birds were by nature seed-eaters

CAROLINA PARAKEET

Psittacidae
Conuropsis carolinensis carolinensis
EXTINCT *c.* 1914
Carolina and Virginia, USA
Conuropsis carolinensis ludoviciana
EXTINCT *c.* 1910
Louisiana, USA

and when the forests were cut to make way for orchards and grainfields, they fed upon these crops as well. The farmers' reaction was predictable.

The Parakeets lived primarily in mature forest areas and formed rookeries within hollow trees. Obviously forest clearance had its effect, but their decline was far more rapid than habitat destruction alone could account for.

There were two easily identifiable subspecies, roughly separated by the Appalachian Mountain ridge that runs north-south down eastern America. The western subspecies (*C. c. ludoviciana*) was generally paler than the eastern type (*C. c. carolinensis*) with a bluer tint to its green colouration and more yellow on the wings. Both races had an instinctive defensive pattern not uncommon in parrots. When one bird had been wounded or killed, the rest of the flock would noisily swoop or hover over the fallen in concern or anger. With certain natural enemies this tactic might distract or drive the predator away from the victim. It was, however, a disastrous practice when confronting a man with a gun, and hunters frequently would destroy entire flocks after bringing down a single bird.

By the 1880s, it became obvious that the birds were extremely rare, but little was done to preserve them until it was far too late. In the east the last wild specimen was collected in 1901, and the last sighting was 1904; although reports of Parakeets being seen in Louisiana were recorded as late as 1910. Others, of course, remained alive in captivity until the last one died in Cincinnati in 1914.

The Carolina Parakeet's fate was untypical for its family, in that its reputation as a pest was largely responsible for its extermination. In contrast, many other species have suffered extinction through human affection or admiration; and the demands of the pet trade still exact an appalling toll on parrots.

This whole family, whose variations of form, size, colour and habitat are enormous – there are over 300 living species – has held a strong fascination for mankind throughout history. Exotic, yet readily tamed; agile, yet content to exercise by climbing in restricted confinement; colourful, playful, mischievous and intelligent, they seem to have been kept as pets by every human society which has evolved in their habitat. But to Europeans, who did not know the birds in any numbers until the colonizing sixteenth century, they are pre-eminently the birds that can talk. Obviously all parrots are not capable of human 'speech', but since classical Greek and Roman times, we have been amused by curious accounts of these talking birds. Without doubt, the most fluent talking bird ever to live was a budgerigar whose owner in Hampshire, England, recorded a vocabulary of 531 words!

BROAD-BILLED PARROT

Psittacidae
Lophopsittacus mauritanicus
EXTINCT c. 1650
Mauritius

To the earliest settlers of the Mascarene Islands, however, anything which moved was meat, and the parrots were not excepted.

As we might expect, the islands were home to a number of highly adapted parrot species, and the scene of almost as many ruthless extinctions. Perhaps the most extraordinary of all parrots once inhabited Mauritius: the *Lophopsittacus* or Broad-billed Parrot which was extinct by 1650.

No specimens of the bird have survived, and our knowledge of this huge parrott (average 70cm, 28in.) might have been restricted to skeletal remains were it not for the discovery in recent times of a detailed sketch by the Dutchman, Wolphart Harmanzoon, in the invaluable manuscript journal of his visit to Mauritius in 1601–02. There is also a rather careless drawing by Sir Thomas Herbert, made in 1638 which may have been the last time the bird was seen alive.

Its main feature was its enormous bill, though skeletal studies suggest that this was a lightweight and weak structure, suited only to a diet of fruit and other soft foods. It was crested, quite long-tailed and bluish-grey in colour. There seems to have been a remarkable difference in size between the sexes, but we can only guess whether the male or the female was the true giant.

But the most significant aspect of the Broad-billed Parrot was its flightlessness. With its great size, reduced keel and very short wings the very most it would have been capable of was a downhill glide like the New Zealand Kakapo's. It was obviously a tempting target for the pot hunters and almost defenceless against the other creatures which men brought with them to Mauritius.

RODRIGUEZ PARROT

Psittacidae
Necropsittacus rodericanus
EXTINCT *c.* 1800
Rodriguez

A possibly related, but far less adapted large parrot inhabited Rodriguez too. In this instance our only evidence is its bones and what can be gleaned from this passage in the anonymous manuscript *Relation de l'Île de Rodrigue* of 1731: 'The perroquets are of three kinds and numerous. The largest are bigger than a pigeon and have a very long tail, and the head, like the bill, is big. Most of them live on islets which lie to the south of the island and where they eat a small black seed produced by a lemon scented little bush, and come to the main island for water. The others remain on the main island where they are found in the smaller trees.'

It is assumed that the largest of these birds was the *Necropsittacus* ('Deadparrot'). On skeletal evidence it was, at 50cm (20in.), the size of a large cockatoo and had a huge bill. There is little more that we can say of it for certain.

However, travellers' descriptions of similar birds on Mauritius and Réunion speak of 'Head and tail fiery red, rest of body and wings green'; and 'Body the size of a large pigeon; head tail and upper part of wings the colour of fire.' Though Rothschild proposed two separate species (*N. francicus* and *N. borbonicus*) based on these descriptions, it is just as likely that we are dealing here with *one* species. On the other hand Count Hachisuka, the Schliemann of Mascarene ornithology, has argued that the Rodriguez bird was probably uniformly green, since the author of the *Relation* goes on to say of the next of his three 'perroquets'; 'The second species is slightly smaller and more beautiful, because they have *green plumage like the preceding*, a little more blue, and *above the wings a little red as well as their beak.*'

Hachisuka's deduction seems fair but must remain conjectural.

RODRIGUEZ RING-NECKED PARAKEET

Psittacidae
Psittacula exsul
EXTINCT *c.* 1880
Rodriguez

'The second species' just quoted from the anonymous *Relation* was almost certainly the Rodriguez Ring-Necked Parakeet (*Psittacula exsul*), so named for its family affinities, not for an actually visible neck ring. It is also known as Newton's Parrot. As recently as 1967 its extinction was regarded as probable rather than certain; but since the last recorded wild sighting was in September 1874 and the last

Mascarene Parrot
Mascarinus mascarinus
Extinct *c.* 1840

Broad-billed Parrot
Lophopsittacus mauritanicus
Extinct *c.* 1650

Rodriguez Ring-necked Parakeet
Psittacula exsul
Extinct *c.* 1880

Seychelles Parakeet
Psittacula wardi
Extinct *c.* 1881

specimen was brought in in August the following year, we must assume that this medium-sized (41cm, 16in.), bluish-green parakeet will not be seen again. It is hard to imagine that the restricted cover remaining on Rodriguez could conceal it for so long.

François Leguat wrote of this parakeet's fondness for the nuts of an oleaceous tree and of the colonists' fondness for the birds' flesh ('not less good than that of young pigeons'); but he also testified to the attraction the exiled Huguenots felt to these parrots as 'delightful pets'. He describes how they trained several of them and took one, which 'spoke French and Flemish', with them when they left Rodriguez for Mauritius.

It would seem from the description in the *Relation* that the adult males of the species might have developed red wing-markings (as they do in related parakeets) but the only specimens in museums today are a female and an immature male.

RÉUNION RING-NECKED PARAKEET

Psittacidae
Psittacula eques
EXTINCT *c.* 1800
Réunion

Closely related to the Rodriguez bird and probably to the just-surviving Mauritian form (*Psittacula echo*, 5 known individuals), was the Réunion Ring-Necked Parakeet (*Psittacula eques*). We know of this bird only from a contemporary plate, published in 1783 which is captioned 'Perruche a collier, de l'isle de Bourbon'; and from the Sieur Dubois' description of 'Green Parakeets, the size of a pigeon, and having a black collar'.

MASCARENE PARROT

Psittacidae
Mascarinus mascarinus
EXTINCT *c.* 1840
Réunion

Another parrot from Réunion is much better documented, though: at least one Mascarene Parrot (*Mascarinus mascarinus*) was brought to Europe and lived out its days in the garden of the King of Bavaria. It was alive in 1834 and its body is presumably the one preserved today in Vienna. None of this species was ever reported in the wild after that date.

Though only about 35cm (14in.) overall it was, like several of the other Mascarene psittaciformes, characterized by a massive beak. It was a strikingly marked and coloured bird, with its lilac head and black face, a red beak and a broad brown tail with a white band at its base. The rest of its plumage was greyish-brown. It is not surprising that it was favoured as a menagerie species but we have no way of guessing how many of them may have set out, at least, from Réunion on the long sea voyage round the Cape.

There is much scholarly disagreement about this parrot's affinities, but it seems likeliest that it was related to the other extinct Mascarene genera and possibly to the scarce black *Coracopsis* of the Seychelles Islands. 'Russet Parrats' having been recorded on Mauritius in 1638 (by Peter Mundy), some writers accept that the Mascarene Parrot may have lived there too. It seems just as likely, though, that Mundy's cryptic reference was to yet another vanished bird from these unfortunate islands.

SEYCHELLES PARAKEET

Psittacidae
Psittacula wardi
EXTINCT *c.* 1881
Seychelles

Further north in the Indian Ocean the birds of the Seychelles Islands fared much better, but there has been a parrot casualty there too, albeit more recently.

The Seychelles Parakeet (*Psittacula wardi*) was related to the Mascarene Ring necks and, like the surviving form in Mauritius, was known as 'Cateau Vert'. Though identical in size (41cm, 16in), it differed from the Mascarene *psittaculas* by its lack of a rosy collar, thus being more closely allied to the Asiatic forms of the genus – though this could be claimed of the Rodriguez bird too. The female bird lacked the partial black necklet of the male but in the field both sexes were characterized by long, pale-blue tails and otherwise green plumage.

On the main island of Mahé, at least, it was regarded as a pest and was both trapped and shot upon sight for its depredations in the maize fields which supported the spreading coconut plantations. It was already very scarce by 1866 when Edward Newton, the British governor of Mauritius, was told on his tour through the Seychelles that it had been exterminated. However, towards the end of his voyage he saw one himself on the small (21sq km, 8sq miles) island of Silhouette, flying warily along the forest edge by the maize fields, and was told that it had once inhabited Praslin Island too.

Four years later we have evidence that there were some Parakeets left on Mahé since a few skins were sent to England (Cambridge); and as late as June 1881 H. M. Warry collected two specimens on the island. This is the last authentic record we have of the Cateau Vert.

CUBAN RED MACAW

Psittacidae
Ara tricolor
EXTINCT *c.* 1864
Cuba

Parrots were known in Europe, though, long before those which came from the islands of the Indian Ocean. Even before the end of the fifteenth century they featured as fruits of the New World in Christopher Columbus's triumphal processions in Spain, and from this period begins the decline of the West Indian (Greater and Lesser Antilles) parrots, and especially the large and spectacular macaws.

The best-authenticated casualty was the Cuban Red Macaw (*Ara tricolor*), not the largest of its kind (only 51cm, 20in.) but gaudily beautiful with its red brow, yellow crown and neck, dark blue wings, and a long tail that was blue above and red below. These macaws nested in holes and clefts in palm trees and favoured those palms and the flowering Melia trees for their diet of fruit, seeds, sprouts and buds. The last recorded wild bird was shot at La Vega in the Zapata Swamp in 1864, though the Paris specimen came from a zoo, probably the *Jardin des Plantes*, and may have been alive later than that.

J. Gundlach collected a number of Cuban Macaws in the 1850s when the last large flock came regularly to feed in a small group of trees at Zarabanda, also in the Zapata Swamp area. Gundlach reported that the Cubans ate the flesh of the Macaw regularly (though he found it repulsive) and that they felled the nesting trees in the hope of capturing undamaged fledglings for pets. (This remains the standard method of collecting for the pet trade in South America.) These depletions, together with the spread of plantations – if, indeed the macaws ever had a wider range than the swamp – eventually extirpated the birds. They were not alone.

Several other West Indian islands harboured macaws which have vanished since European contact, through the combined demands of plantation expansion and the exotica trade. But only the Cuban Red Macaw is represented by specimens in modern collections and a brief review of the evidence for other members of its family must suffice.

That the Carib Indians, themselves casualties of European expansion, ate and domesticated macaws and parrots on various Antillean islands we know from the testimony of Christopher and Ferdinand

Columbus. One of these birds has left only a single leg-bone to posterity: it was found in a Carib or Arawak kitchen-midden on St Croix, St Vincent. This has been named *Ara autocthenes*, and cannot of course be even tentatively illustrated here.

Also unpictured, but better documented, are the Guadeloupean and Hispaniolan relatives of the Red Macaw, called 'Guacamayo' by the Caribs as the Cuban bird was 300 years later. In April 1496 Ferdinand Columbus reported seeing 'red parrots as large as chickens' on Guadeloupe; and de las Casas, in his history of the Indies, differentiated the Hispaniolan Macaw from the one on Cuba by its white instead of yellow brow. Moreover, we have seventeenth-century pictorial evidence from Roelandt Savery (the chronicler of the Dodo) of a macaw which exactly fits this description. Whether or not we should assume *two* other red macaws from those records is a matter for guesswork.

YELLOW-HEADED MACAW

Psittacidae
Ara gossei
EXTINCT *c.* 1765
Jamaica

It is also not impossible that the Jamaican Yellow-Headed Macaw, (*Ara gossei*) – last seen about 1765 at Lucea near Montego Bay – was the same as one or both of the above parrots. The stuffed but legless body of this bird was seen by a Dr Robinson who described it in sufficient detail to the naturalist Gosse for us to illustrate it with some confidence.

GREEN AND YELLOW MACAW

Psittacidae
Ara erythrocephala
EXTINCT *c.* 1842
Jamaica

A somewhat later survivor in Jamaica was seen by the Reverend Comard in 1842 in the parish of St James near the island's heart. He observed two large macaws flying near the foot of the mountains and was told by residents that they were, from beneath, vividly yellow and blue plumaged. Almost certainly (though in this conjectural field some authorities will disagree) these were the same species as the bird procured in 1810 in the mountains of Trelawney and St Anne's by the proprietor of the Oxford Estate, Mr White.

One of Gosse's 'ornithological acquaintances', Mr Hill, who believed that these macaws wintered in Jamaica from Mexico, described it thus: 'Head red; neck, shoulders and underparts of a light and lively green – the greater wing coverts and quills blue; and the tail scarlet and blue on the upper surface, with the under plumage both of wings and tail a mass of intense orange yellow.'

This description seems detailed enough for us to include a tentative reconstruction of the Green and Yellow Macaw *(Ara erythrocephala).*

DOMINICAN MACAW

Psittacidae
Ara atwoodi
EXTINCT *c.* 1800
Dominica, West Indies

Cautiously, we include also an illustration of the macaw from Dominica, *Ara atwoodi*, described by Thomas Atwood in his 1791 account of that island. In his words 'The Mackaw is of the parrot kind, but larger than the common parrot and makes a more disagreeable noise. They are in great plenty, as are also parrots on the island: have both of them a delightful green and yellow plumage, with a scarlet coloured fleshy substance from the ears to the root of the bill, of which colour is likewise the chief feathers of the wings and tail.' The 'great plenty' of these striking parrots was evidently short lived: there are no later accounts of them and no known specimens in existence. Still one of the most exploited groups of birds, the Macaws may benefit most from the IUCN's successful bid to give, in 1981, protection to all but three of the world's 380 parrot species.

Cuban Red Macaw *Ara tricolor* Extinct *c.* 1864

LABAT'S CONURE

Psittacidae
Aratinga labati
EXTINCT *c.* 1722
Guadeloupe, West Indies

Spectacular as the Macaws were, their smaller relatives, the Conures (*Aratingae*) and Amazons proved just as attractive to the exotica trade, and equally delicious in the pot and unwanted on the plantations. Though there are Conures today in Cuba and Jamaica, and *Aratinga chloroptera* survives comfortably on Hispaniola, a subspecies (*A. c. maugei*) was extirpated from Puerto Rico before the end of the last century and 150 years earlier the Guadeloupean form (*A. labati*) disappeared.

Conures are small to medium-sized parrots (average 32cm, 12.5in.), with long, gradated tails, naked eye-rings and broad, heavy bills. We know of the Guadeloupe form, *Aratinga labati*, from the 1722 *Nouveau Voyage* . . . by J. B. Labat for whom the bird is named and who wrote: 'Those of Guadeloupe are about the size of a blackbird, entirely green, except for a few small red feathers on their heads.' Earlier Du Tertre in his 1667 *Histoire generale* . . had distinguished the Conures from the Aras (Macaws) and 'perroquets' (Amazons) by saying: 'Those which we call Perriques are the small Perroquets, green all over and the size of Magpies.'

PUERTO RICAN CONURE

Psittacidae
Aratinga chloroptera maugei
EXTINCT *c.* 1892
Mona Island, Puerto Rico

A more recent casualty was the small subspecies of the Hispaniolan Conure, *A. c. maugei*, which lived on Mona Island, midway between Hispaniola and Puerto Rico, and – according to oral tradition among old-timers – on Puerto Rico itself.

A duller green bird than the Hispaniolan form, with more extensive red markings on the underwing coverts and a darker, smaller bill, this Conure was last collected in 1892 by W. W. Brown, having been drastically reduced through the last century by visiting pigeon hunters.

These were highly gregarious parrots, characterized by non-stop screeching as they pursued their regular 'flight paths'. Within a group, each pair formed a distinct unit, and the flocks moved swiftly and directly. They fed on seeds, fruits, nuts, berries and probably leaf buds and blossoms. Though normally wary birds, their caution seemed to vanish when they were feeding, and since they adapted to feeding in maize fields, their destruction is the more readily understandable.

They nested in hollow trees, old woodpecker holes and arboreal termite nests.

By one of the ironies of the pet-trade, their range has now been taken over by the Hispaniolan Amazon (*Amazona ventralis*) after a consignment of those birds was refused entry by the Puerto Rican authorities and released at sea.

GUADELOUPE AMAZON

Psittacidae
Amazona violacea
EXTINCT *c.* 1750
Guadeloupe, West Indies

The parrots of the genus *Amazona* have suffered heavy losses, and most surviving species are rare or endangered. Characterized by their duck-like flight, with shallow wing beats below their body-lines, these large, colourful parrots, with their strong, heavy beaks and short, slightly rounded tails are an extraordinary group. They are representatives of a very old branch of the Parrots and their relationship to other genera is unclear. James Greenway writes: 'The most surprising thing about them is that such specialised birds, being on small islands and surrounded by enemies, should still exist.'

Both the Guadeloupe and Martinique Amazons are long extinct, however, and though the Dominican form (the spectacular 'sisserou',

Labat's Conure
Aratinga labati
Extinct *c.* 1722

Green and Yellow Macaw
Ara erythrocephala
Extinct *c.* 1842

Puerto Rican Conure
Aratinga chloroptera maugei
Extinct *c.* 1892

Dominican Macaw
Ara atwoodi
Extinct *c.* 1800

Yellow-headed Macaw
Ara gossei
Extinct *c.* 1765

A. imperialis) was preserved until 1979 by the mountainous nature of that island's forests, there is no knowing yet what effect the cataclysmic hurricane of that year may have had upon it.

Once again we have Du Tertre's and Labat's observations to thank for our knowledge of the Guadeloupe birds, *Amazona violacea*. Du Tertre reported that the French settlers hunted the Amazons for food as, probably, did the slaves imported from Africa. The almost total forest clearances for plantation and the dense population (by 1900 there were 208 people per sq km, 540 per sq mile) must have been the main causes of their disappearance, though.

Besides, it is easy to imagine the appeal of these parrots to the exotica merchants. Du Tertre's detailed description is vivid: '... about the size of a hen. It's beak and eyes are rimmed with red, while the feathers of the head, neck and belly are violet mingled somewhat with green and black and iridescent like a pigeon's throat. The whole back is green with a marked brown [tinge]. The three or four chief wing feathers are black and all the rest yellow, green and red. It has two beautiful "roses" of the same colours on the main body of the wings. When it ruffles its back feathers it seems to create a frilled collar round its head.'

Labat's account of the Guadeloupe Amazon stresses grey rather than violet as the parrot's ground colour, but it may well be that he confused this bird with the Martinique form since he describes them as identical.

MARTINIQUE AMAZON

Psittacidae
Amazona martinica
EXTINCT *c.* 1750
Martinique, West Indies

There certainly was a species of Amazon on Martinique and, just as certainly, this parrot (*Amazona martinica*) fell victim to the same pressures as its Guadeloupean congener. Like Guadeloupe, Martinique was suitable for cultivation and was heavily cleared and populated (185 people per sq km, 480 per sq mile by 1900). Labat's is the only description known, and forms the basis of our reconstruction.

CULEBRA ISLAND AMAZON

Psittacidae
Amazona vittata gracileps
EXTINCT *c.* 1899
Culebra Island, Puerto Rico

Human progress took longer to oust the Culebra Island Amazon, (*A. vittata gracileps*). It was last seen by A. B. Baker who shot three of them in 1899. It was a smaller, lighter-footed form of the Puerto Rican Amazona (*A. vittata*) and some authorities regard it as too poorly differentiated to be regarded as a full sub-species. However, *A. vittata* is itself in grave peril today. In 1977 there were only 15–20 wild individuals left – among which there were at most five breeding pairs – and 14 captives. The future of the species is obviously very doubtful.

The Culebra Island Amazons were a green form, highly gregarious (except near the nest where they were 'viciously territorial'), following regular flightpaths morning and evening and avoiding the clouds by circling the mountain peaks.

Their diet originally included at least 50 different fruits but especially those of the Sierra Palm (in the breeding season) and the Tabunoco (in the autumn). They also took nectar and on one occasion a bird was seen carrying off a lizard (*Anolis cuvieri*).

The attraction that the Amazons had for the pet trade was probably enhanced by their individualistic nature. Their calls are varied and complex and even within a flock each individual has its own 'voice'. We can only hope that the authorities' efforts will prevent those distinctive voices from being stilled forever.

Guadeloupe Amazon *Amazona violacea* Extinct *c.* 1750

NEW CALEDONIAN LORIKEET

Psittacidae
Charmosyna (Vini) diadema
EXTINCT *c.* 1860
New Caledonia, Southwest Pacific

After the Mascarenes and the Antilles we turn inevitably to that other theatre of historical European expansion – the Pacific and the Antipodes.

A whole family of parrots, the Lories and Lorikeets, is at risk today throughout the Pacific though only one form, the New Caledonian Lorikeet (*Charmosyna diadema*), has so far been reported actually extinct.

These small (16–20cm, 6½–8in.), elegant birds have suffered from forest destruction, shooting, collectors' demands, and a host of introduced animals; but the most serious, and recent, threat comes with the advent of modern international air transport to the Pacific islands. By this agency, species of mosquito have been introduced which are carriers of avian malaria.

The New Caledonian Lorikeet is known only from two female specimens (18cm, 7in.) collected before 1860, but even these old skins retain a spectral beauty.

The female was yellow at cheek and throat with a crown of violet-blue and a deep-orange bill. The main body was green, but the thighs were blue-tinged and there were red and black markings near the vent and tail. The narrow, gradated tail itself was green above and yellow below. The legs were orange.

It was a prize calculated to excite any aviarist or collector. Unhappily, this is still the case and officials have complained that New Caledonian natives are on the lookout for the Lorikeets after rumours of their rediscovery. They have been offered enormous sums by collectors for specimens dead or alive.

It makes no difference to a bird population, of course, whether men

Martinique Amazon
Amazona martinica
Extinct *c.* 1750

Culebra Island Amazon
Amazona vittata graciliceps
Extinct *c.* 1899

do take specimens dead or alive. In either case a wild breeding pair may be broken up, and only in specialized hands will captive birds reproduce. More commonly they are kept as pets or exotics in isolation, and all too often the last representative of a species dies in a small cage, mourned by some human family no more than if it had been a budgie or a goldfish.

Such was the fate of the Norfolk Island Parrot, a large, beautiful and intelligent Kaka, about which we know far more in its role as family pet than as a wild creature.

Like most birds on this convict island, the Kaka got short shrift from the unwilling settlers. John Gould, our only real authority, wrote prophetically in 1841: 'the native haunts of this fine bird have been so intruded upon, and such a war of extermination been carried out against it, that if such be not the case already, the time is not far distant when . . . like the Dodo, its skin and bones become the only mementos of its existence.' Its last stronghold was Phillip Island, 4.8km (3 miles) from the main island and only 8km (5 miles) in circumference.

The Kaka seems to have been a forest dweller, like its very much larger (45cm, 17½in. against 38cm, 15in.) relative in New Zealand, and to have lived primarily on the nectar of the White Hibiscus or Whitewood Tree. Gould examined the tongue of a pet bird and found 'a very peculiar structure', not brush-ended like the nectar-feeding parakeets' but 'with a narrow horny scoop on the underside which, together with the extremity of the tongue, resembled the end of a finger with the nail beneath instead of above.' He was told of one pet

NORFOLK ISLAND KAKA

Psittacidae
Nestor productus
EXTINCT *c.* 1851
Norfolk and Phillip Islands, New Zealand

New Caledonian Lorikeet
Charmosyna (Vini) diadema
Extinct *c.* 1860

lk Island Kaka
or productus
inct *c.* 1851

Macquarie Island Parakeet
Cyanoramphus novaezelandiae erythrotis
Extinct *c.* 1890

bird which 'evinced a strong partiality to the leaves of the common lettuce and other soft vegetables and was also very fond of the juice of fruits, of cream and of butter.'

The Kaka on which Gould made his own observations was the household pet of Major and Mrs Anderson in Sydney, Australia. It was not caged but 'permitted to range over the house, along the floors of which it passed, not with the awkward waddling gait of a Parrot, but in a succession of leaps, precisely after the manner of the Corvidae.'

Gould was fascinated by the pet, noting its 'hoarse, quacking inharmonious voice, sometimes resembling the bark of a dog', and he found its behaviour and actions so unusual for a Parrot that he was 'convinced that they were equally different and curious in a state of nature'. All he could discover about the wild birds, though, was gleaned from Mrs Anderson who told him that on Phillip Island they frequented rocks as well as the treetops, and nested either in holes or in hollow trees, laying as many as four eggs.

Gould concluded that the Kaka 'bears captivity extremely well, readily becoming contented, cheerful and an amusing companion'. One such bird, almost certainly the last of its kind on earth, died in a cage in London some time after 1851.

MACQUARIE ISLAND PARAKEET

Psittacidae
Cyanoramphus novaezelandiae erythrotis
EXTINCT *c.* 1890
Macquarie Island,
New Zealand, Southern Ocean

The most far-ranging of all the Parrot families, the *Cyanoramphus* Parakeets, once had a range of at least 55° longitude and 30° latitude and comprised six species, one (*novaezelandiae*) with nine distinct subspecies. These small to medium (average 26cm, 10in.), stocky birds with long, gradated tails are extremely adaptable and only those from smallish islands have been lost, though we may be fairly sure that there have been some unrecorded extinctions. The New Zealand forms of the *Cyanoramphi* demonstrate the genus' adaptability, inhabiting such varied regions as the forest reserves on New Zealand itself, and the gale-ripped treeless island of Macquarie far to the south.

But though the Macquarie Island Parakeet (*C. n. erythrotis*) adjusted expertly to an inhospitable, terrestrial environment it did not survive the coming of sealers and penguin-hunters or, rather, the cats which they left behind them.

In 1820 a Russian expedition under Bellinghausen visited Macquarie, collected 20 specimens of the parakeet, and procured one live bird from a sealer for three bottles of rum. From that time there was a fairly regular trade in living specimens which were sold in Sydney as good cage birds.

In 1880 J. H. Scott of the University of Otago went to Macquarie, described the birds nesting under the tussock grasses, and said that they could be seen 'in great numbers round the shore'. But only 14 years later A. Hamilton, from the same University, could not find any. There is little doubt that abandoned cats, rather than the pet-trade, wiped out the tough little parakeets.

Red-fronted Parakeet
Cyanoramphus novaezelandiae subflavescens

RED-FRONTED PARAKEET

Psittacidae
Cyanoramphus novaezelandiae subflavescens
EXTINCT *c.* 1869
Lord Howe Island, Tasman Sea

At about the same time the Lord Howe Island Red-fronted Parakeet (*C. n. subflavescens*) met its end. Though it had survived the settlement by Europeans for over 50 years it was regarded by them strictly as a pest. Like the Macquarie Island form it was relatively large (27cm, 10½in.) and it was distinguished too by a much less red head and more generally yellow plumage than the New Zealand birds'. This parakeet was last reported by E. S. Hill when he accompanied a judicial party to Lord Howe Island in 1869. In his published notes the next year he wrote: 'The parraquet also was a nuisance to the cultivators, once appearing in flocks; now I saw but a solitary pair in their rapid flight through the air and recognised them only by their peculiar noise.'

He was probably the last man without a gun to hear them.

BLACK-FRONTED PARAKEET

Psittacidae
Cyanoramphus zealandicus
EXTINCT *c.* 1850
Tahiti

The most isolated examples of this genus were found on the Society Islands, 3200km (2000 miles) northeast of their main concentration round New Zealand. Two species, unusually dark-coloured, inhabited Tahiti and Raiatea. The largely dark brown Society Parakeet (*C. ulietanus*) has, astonishingly, been reported rediscovered after more than 200 years. Assuming this is true, it is a heartening indication of what a patch of wilderness and a healthy dose of human indifference can accomplish for conservation.

The Tahiti Black-fronted Parakeet (*Cyanoramphus zealandicus*) was basically green, like most *Cyanoramphi*, with some blue markings and a scarlet stripe behind the eyes, but it was strikingly different in its sooty black brow. It was probably uncommon even when first collected in 1773, two specimens being brought home from Cook's second voyage, and presumably became extinct shortly after 1844 when a French serviceman, Lieutenant de Marolles, brought one back to Paris. The natives called it the 'Aa', as Parkinson, one of Cook's naturalists, noted; and today that name is the only remembrance of the parakeet that the Tahitians retain.

Black-fronted Parakeet *Cyanoramphus zealandicus*

White Gallinule *Porphyrio albus* Extinct *c.* 1830

Walking Birds

THE RAILS AND THEIR ALLIES

The Rails form one of the most ancient and widespread bird families (Rallidae) and have their origins in the tertiary period of 70 million years ago. Varying from the size of a robin to that of a goose, they can be found everywhere but the polar regions.

Rails are distinguished by the narrowness of their bodies, adapted to swift running through underbrush. The expression 'thin as a rail' originates from these birds, not from an iron or wooden bar. They are generally secretive and well-camouflaged birds, with bills and feet of widely diversified shapes to adapt to specific terrains. Though generally weak fliers, rails have colonized all the major oceanic islands and many of the smallest and most remote ones. Many have been so long isolated they have developed monotypic genera, and a large percentage have become flightless.

In layman's terms the Rallidae family is made up of rails, crakes, coots and gallinules; and belongs to the order of birds known as Gruiformes, which includes bustards and cranes. This order, broadly defined, is made up of ground-feeding, ground-nesting, walking birds that rarely fly, and many of their number (especially the cranes) are today critically endangered.

WHITE GALLINULE
Rallidae
Porphyrio albus
EXTINCT *c.* 1830
Lord Howe Island, Tasman Sea

Not all rails were wary and camouflaged birds. One of the largest and most striking was the White Gallinule (*Porphyrio albus*) of Lord Howe Island. This red-billed, white-feathered and flightless bird, which is also called the White Swamp-Hen, is known from only two existing skin specimens, one written account, and two illustrations by Thomas Watling and George Raper from 1788 and 1790.

The White Gallinule did not long survive the coming of men. It was likely killed off even before the first small settlement of 1834, and a thorough description of the island's avifauna made in 1847 specifically states that there were no 'white fowles'. Almost certainly, as the British officers stated how easily these birds were run down and killed with sticks on the tiny island, it was whalers, naval crews and convict supply ships which extirpated this unique creature.

WAKE ISLAND RAIL
Rallidae
Gallirallus wakensis
EXTINCT *c.* 1945
Wake Island, North Pacific

Wake Island Rail
Gallirallus wakensis

If one were to make a broad study of the rails of the world, it would necessarily require an island-hopping journey around the entire globe. Perhaps the best place to begin would be the tiny mid-Pacific island where the most recently extinct rail, the Wake Island Rail (*Gallirallus wakensis*), perished.

The Wake Island Rail's history had a gruesomely sad and bizarre end. Having survived the most dangerous early period of human contact it became a direct casualty of World War II. In the great war of the Pacific, Wake Island became a Japanese garrison. The island was only 6m (20ft) above sea level and covered with Pandanus scrub. It supported no other land birds at all. As the war continued, the plight of the Japanese, isolated and cut off from supplies, became desperate. The soldiers were starving. One by one, every rail on the island was hunted down and eaten.

This bird was a small and flightless descendent of the Philippine Banded Rail. Its narrow white bars on breast, abdomen and flanks were striking, as was the grey stripe from the bill over the top of its eye. It had been isolated from its parent group in the Philippines for long enough to have evolved into 'an extreme variant'.

The most accurate notes on the rails were made by Dr Alexander Wetmore during his 1923 visit. He observed them in their breeding season between July and August and commented on their alertness and curiosity as they came out from cover 'with head and neck erect and jerking tails'. It could not have been too difficult for the Japanese to catch them, though Wetmore did say that they would 'skip rapidly' back into cover at any sharp movement.

According to pre-war visitors to Wake Island this flightless rail was common, unafraid and without significant enemies. In the context of Hiroshima, its passing went for several years unnoticed.

IWO JIMA RAIL
Rallidae
Poliolimnas cinereus brevipes
EXTINCT *c.* 1924
Iwo Jima, North Pacific

Another scene of severe fighting in the North Pacific was Iwo Jima or Sulphur Island, which had been the home until 1924 of another rail, known to the Japanese as the 'Mamjiro Kuina' (*Poliolimnas cinereus brevipes*). A small (15.3cm, 6in.) bird, dark brown and mottled with black above and white below, it succumbed to the clearing of the land for sugarcane and to the epidemic of rats. Previously the only mammals on the island had been bats.

The clearance of the rich natural forest reduced the availability of water, so that by the turn of the century this subspecies of the widespread Ashy Crake was obliged to come down to the human settlements in dry weather to drink at water tanks. Here the birds very often fell victim to cats. The Iwo Jima Rail may have been extinct since 1911 when the last specimen was taken – but the Japanese naturalist, T. T. Moniyama, did report sightings in 1924.

KITTLITZ'S RAIL
Rallidae
Porzana monasa
EXTINCT *c.* 1850
Kusaie and Ponape, Caroline Islands, North Pacific

Further south, in the Caroline Group, are the islands of Kusaie and Ponape, where F. H. von Kittlitz collected two specimens of a small black rail about 17.8cm (7in.) long in December 1827. He noted that its beak was larger and heavier than its relatives in mainland Asia; and that only its red feet and eyes showed any vividness. It lived singly on the ground, in a variety of habitats: swamps, saltmarshes and 'wet shadowy places in the forest'.

Kittlitz's Rail (*Porzana monasa*) or the 'Ponape Crake' as it also became known was certainly gone from the islands within 50 years of Kittlitz's visit, though to this day it remains a legend among the natives, since it had been a sacred bird before the onslaughts of missionaries upon the indigenous religion. Its names were 'Satamanot' and 'nay-tai-mai-not', or 'one who lands in the Taro plot' – and the latter name might suggest that it still had some powers of flight.

The basic cause of the bird's extinction is not in question: Kusaie was overrun by rats from the time when whalers began to put in and 'heave down' their ships in the 1830s and 40s. All that remains of it are two specimens in Leningrad, and Kittlitz's descriptions, which even in translation retain a poetic wistfulness: 'One hears in these places from time to time its alluring voice resounding.'

Kittlitz's Rail
Porzana monasa

FIJI BARRED-WING RAIL
Rallidae
Nesoclopeus poeciloptera
EXTINCT *c.* 1965?
Fiji Islands

A much larger member of the rail family, the Barred-wing Rail (*Nesoclopeus poeciloptera*) or 'Mbidi' of the Fiji Islands, was a good 61cm (24in.) in length, and not at all the drab creature of the Carolines yet, according to E. L. Layard's notes of 1875, it was every bit as shy and elusive. It stayed hidden at all times in swamps, and nested among the reeds. Although the wings of specimens seem capable of flight, both the Fijians and Europeans observers insisted that it never left the ground.

It has been widely assumed that introduced mongooses exterminated it from its native islands of Viti Levu and Ovalau before 1890, but these are steep and forested places and more recent, though unconfirmed, sightings have been reported.

SAMOAN WOOD RAIL
Rallidae
Pareudiastes pacificus
EXTINCT *c.* 1873
Savaii, Samoa, Central Pacific

A little to the east, on the Samoan island of Savaii, lived the Wood Rail (*Pareudiastes pacificus*) or Punae, a small (15.3cm, 6in.) black gallinule, practically flightless and rendered even more elusive than most small Pacific rails by its nocturnal habits. It had enormous eyes and ate insects and, other than its inability to adjust to captivity, we know very little more about it. It was last recorded in 1873 when the *Challenger* Expedition took two specimens.

TAHITI RAIL
Rallidae
Gallirallus ecaudata
EXTINCT *c.* 1900
Society Islands

Still less is known about the brilliantly coloured 23cm (9in.) red-billed Tahiti Rail (*Gallirallus ecaudata*) of the Society Islands, which was never collected and which is only known to us through the (certainly reliable) plate by Captain Cook's ornithological illustrator, G. R. Forster.

The native Polynesians called it the Tevea or Eboonaa though they have recorded nothing about its habits; but there is little mystery about its disappearance. Even the smallest of the Society Islands are overrun by rats – though cats may have been the worst predators, since the bird was reported to have survived on Mehetia (which has no cats) until the turn of this century.

MODEST RAIL

Rallidae
Gallirallus modestus
EXTINCT *c.* 1900
Chatham Islands, New Zealand

Modest Rail
Gallirallus modestus

Off New Zealand's east coast, the Chatham Island group was the home of two other members of the Rail family, now extinct. The smaller of these, the Maori 'Matirakahu' (*Gallirallus modestus*), was the first to colonize the group but was already struggling to hold its own when Europeans arrived. This little (17.8cm, 7in.) rail was pale brown with a lightly barred breast and showed, for its family, a remarkable sexual dimorphism: the female was more strongly barred and had a much shorter bill. The 'Modest Rail' had so limited an ecological niche that even the sexes had to adopt somewhat different feeding habits.

It had, up to the nineteenth century, occupied both Pitt and Chatham Island in the group but when it was discovered by H. H. Travers in 1871 it only survived on tiny Mangere Island (1.25sq km, .43sq miles). The most probable reason for its extirpation elsewhere was probably not man or predators, but competition from a *parvenu* relative, the Dieffenbach Rail. S. D. Ripley makes an analogy with the earlier, smaller Polynesians giving way to the later invasion of the Maoris. This in turn leads him to compare the rats and cats which came eventually to Mangere to Europeans – as he says, 'a mournful thought indeed'.

As to its fate in its last stronghold, the New Zealand authority W. R. B. Oliver says bluntly: 'Through the work of collectors, in order to gain profit the Chatham Island Rail was exterminated about 25 years after it was discovered'; but other writers have pointed out that bush fires and the introduction of goats and rabbits, meant a total destruction of its habitat anyway.

Of its habits we know only that it nested in holes or burrows, that its young commonly sheltered in hollow logs and that it lived on insects, principally beetles and the sand-hoppers (crustaceae) which travelled some way in from the shore.

DIEFFENBACH'S RAIL

Rallidae
Gallirallus dieffenbachii
EXTINCT *c.* 1840
Chatham Islands, New Zealand

The second Chatham Island Rail was Dieffenbach's Rail. Although this larger (28cm, 11in.), strongly-marked bird had ousted its predecessor from Chatham Island, it did not survive as long in the world. The only specimen collected was taken in 1840.

Its discoverer, E. Dieffenbach, wrote in his 1843 book *Travels in New Zealand*: 'It was formerly very common, but since cats and dogs have been introduced it has become very scarce. The natives call the bird 'Meriki', and catch it with nooses. I often heard its shrill voice in the bush, and after much trouble obtained a living specimen.'

The bird had adapted to life in the cover of tussock grass and to probe-feeding (hence its distinctive beak tip), and nested among the tussocks. To Dieffenbach's mention of cats and dogs must be added the ubiquitous rats and, probably most decisive, the common occurence of bush fires once the Polynesians had settled.

Far to the south of New Zealand (lat. 54°45S) is the 260 sq km (100sq mile) island of Macquarie, a windswept spot where only tussock grass and mosses thrive. Its endemic birds (a rail and a parrot) are both extinct, though exotics have been introduced.

The Macquarie Island Rail (*Gallirallus philippensis macquariensis*) has been classed by many scientists as identical to *G. p. assimilis*, the Banded Land Rail of New Zealand, but S. D. Ripley, at any rate, distinguishes it by its short bill, darker wings and the strong rufous-coloured band on its chest, and W. R. B. Oliver also points to its stouter legs.

The sealing company of Elder and Cormach brought a live specimen to New Zealand in March 1879 and reported that it was common at the south end of the island. The next year J. H. Scott went to Macquarie and said they were 'not at all common. There seemed to be two varieties – one, slightly the larger, was reddish in colour, the other was black' (he may well have seen immatures which *are* black). Fourteen years later, A. Hamilton 'did not see any birds like the banded rail' though he saw plenty of rats.

MACQUARIE ISLAND BANDED RAIL
Rallidae
Gallirallus philippensis macquariensis
EXTINCT *c.* 1880
Macquarie Island, New Zealand, Southern Ocean

Moving east into the Indian Ocean we encounter once more that familiar disaster area, the Mascarenes. There were certainly flightless rails on these islands and there are numerous descriptions and some bone remains. But we can only be reasonably sure of the conformation, size and colour of two of the five or six reported species which rapidly ended up in the cooking pots of settlers and seamen.

For the Mauritian Red Rail (*Aphanapteryx bonasia*) there are five contemporary illustrations – three painted from specimens in

MAURITIAN RED RAIL
Rallidae
Aphanapteryx bonasia
EXTINCT *c.* 1680
Mauritius

Samoan Wood Rail
Pareudiastes pacificus

Tahiti Rail
Gallirallus ecaudata

Iwo Jima Rail
Poliolimnas cinereus brevipes

Fiji Barred-wing Rail
Nesoclopeus poeciloptera

European zoos by Roelandt Savery – as well as descriptions and bones. All the pictures show a large rail, long-legged, short-winged and, with some variations, of a reddish-brown colour.

In 1638 François Cauche wrote of: 'red hens with the beak of a woodcock . . . to capture them one need only show them a scrap of red cloth which they pursue and let themselves be seized in your hand; they are the size of our hens and excellent eating.'

In the same year the diarist Peter Mundy wrote of 'yellowish wheaten coloured hens on Mauritius'. Some writers have assumed that this was another rail, others that it was in juvenile plumage. Mundy commented on their flightlessness and mentioned a 'pretty way of taking them' with a red cap.

In 1673 this was still apparently a popular sport. Johann Hoffman wrote then of holding a red cloth in the left hand and a rod in the right to strike the 'hens' with as they approached. He also commented on the speed with which they could run and their long curved beak. The cries of a wounded bird, apparently, would attract others. This easy method of hunting seems to have wiped out all the birds soon after – they were not reported again.

LEGUAT'S RAIL

Rallidae
Aphanapteryx leguati
EXTINCT c. 1700
Rodriguez

The isle of Réunion was neither exploited nor reported on so thoroughly in the seventeenth century, though we may assume that there were rails there; but we do have acceptable documentation of one species on Rodriguez, thanks again to the pen of François Leguat.

In 1691 he wrote: 'Our woodhens (Gellinotes) are fat all the year round and of a most delicate taste; they are always coloured bright grey.' He described the adults as having a red ring around each eye and straight, sharp beaks about two inches long and also red.

Leguat's Rail (*Aphanapteryx leguati*) seemed to share with its relatives a fascination with the colour red: 'If you offer them anything red they get so angry that they will dart at you to snatch it out of your hand and in the heat of combat we can take them with ease.' And Leguat states unequivocally: 'They cannot fly; their fatness renders them too heavy for that.'

The author of the *Relation de l'Ile de Rodrigue* described the birds as continually whistling except when pursued, at which time 'they produce another sort of call which one might compare to the sound of a man with hiccups'.

TRISTAN GALLINULE OR ISLAND HEN

Rallidae
Gallinula nesiotis
EXTINCT c. 1890
Tristan da Cunha, South Atlantic

East again to the South Atlantic where Tristan da Cunha had its rails too. Gough Island still has its *Gallinula comeri*, but the rail (*Gallinula nesiotis*) from Tristan itself has been extinct since the late 1880s. In 1882 the *Henry B. Paul* was wrecked there and rats came ashore. The already scarce birds were doomed.

The rail was first described scientifically by Sclater in 1861 but in 1811 Jonathan Lambert, self-styled King of Tristan, wrote: 'We have the little black cock in great numbers and in the fall (they) are very fat and delicate. We caught some hundreds last year with a dog.'

August Earle, future draughtsman of the *Beagle* and the sadly comical victim of a captain who took off in bad weather and marooned him for six months, studied the bird closely: '. . . about the size of a partridge, but their gait was something like that of a penguin. The male is of a glossy black with a bright red hard crest on top of the

Dieffenbach's Rail
Gallirallus dieffenbachii
Extinct *c.* 1840

Tristan Gallinule
Gallinula nesiotis
Extinct *c.* 1890

Macquarie Island Banded Rail
Gallirallus philippensis macquariensis
Extinct *c.* 1880

Mauritian Red Rail
Aphanapteryx bonasia
Extinct *c.* 1680

Leguat's Rail
Aphanapteryx leguati
Extinct *c.* 1700

head. They stand erect and have long yellow legs with which they run very fast; the wings are small and useless for flying, but they are armed with sharp spurs for defence and also, I imagine, for assisting them in climbing, as they are generally found among the rocks. The name they give this bird is simply 'Cock' its only note being a noise very much resembling the repetition of that word. Its flesh is plump, fat and excellent eating.'

JAMAICAN WOOD RAIL OR UNIFORM RAIL

Rallidae
Aramides concolor concolor
EXTINCT *c.* 1881
Jamaica

There were certainly two rails on St Helena, one of which, the Little St Helena Rail, *Porzana astrictocarpus*, was living in 1502, but there is skeletal evidence only from which to describe it. The last Atlantic casualty of the family was the Jamaican Wood Rail or Uniform Rail (*Aramides concolor concolor*).

Like so many Jamaican species this largish (25.5cm, 10in.) reddish-brown rail had rats, cats and, above all, mongooses to contend with. Last collected in 1881, it must have died out soon afterwards. It was a versatile bird, inhabiting inland swamps and streams, but also fairly high altitudes. It could fly, though apparently in a laboured manner, and always preferred to run.

P. H. Gosse, who named it, wrote 'It is sometimes seen perched on a low tree by the roadside, at which time it seems to have lost its normal shyness, and sits looking at the sportsman until he nearly comes up to it . . . I have shot it skulking among the aquatic weeds at Basin Spring. As it roams, it utters at intervals of a few seconds, a cluck, like a hen . . .'

Sandwich Rail
Porzana sandwichensis

SANDWICH RAIL

Rallidae
Porzana sandwichensis
EXTINCT *c.* 1884
Hawaii

For the last two Rail casualties we come full circle around the globe to the Hawaiian Islands. On the eastern half of Hawaii itself, there once lived the minute (14cm, 5½in.) brown crake known as the Sandwich Rail (*Porzana sandwichensis*). This bird preferred the open grassy areas just below the heavy forest line. It was flightless, though a fast runner. Its local name was Moho: 'bird that crows in grass'. R. C. L. Perkins – the first real authority on Hawaiian vertebrates – comments on its habit of taking refuge in the holes of the non-carnivorous native rats. It would be ironic if the rails had continued this tactic once the European rats had arrived – it would certainly have hastened their extinction.

Certainly rats, cats and dogs are far more likely villains, for once, than the mongoose, since the viverrine was not introduced until 1883, only a year before the rail's last sighting.

LAYSAN RAIL OR SPOTLESS CRAKE

Rallidae
Porzana palmeri
EXTINCT *c.* 1944
Laysan, Hawaii

The related Laysan Rail (*Porzana palmeri*) is a much better-documented bird and a much sadder and more recent extinction. For Man, who was so much responsible for its loss, had come very close to redeeming himself by saving it.

Laysan is a very small (5.2sq km, 2sq miles), low island towards the western end of the chain, and first reached by Europeans in 1828. The small (15.3cm, 6in.) crake was common then and fearless, and does not seem to have been directly molested by the guano diggers, but in 1903 rabbits were introduced and by 1923 had totally destroyed the grass patches and scaevola thickets in which the birds lived and nested. By the mid-twenties they were all gone from Laysan but survived on the tiny islets of Sand (3.9sq km, 1.5sq miles) and Eastern (2.6sq km, 1sq mile) on the Midway Atoll to which they had been introduced in 1891 and 1910 respectively.

In 1892 Frohawk wrote: 'Soon after dusk they all, as if by one given signal, strike up a most peculiar chorus, which lasts but a few seconds, and then all remain silent. I can only compare this song to a handful or two of marbles being thrown on a glass roof and then descending in a succession of bounds.'

Their fearlessness, inquisitiveness and swiftness ('a shadow with mouse-like speed') were commented on by all who encountered them. They were known to hop on to the laps of US sailors stationed on Midway during the war and to forage for crumbs around the dinner tables in the mess; and they were treated, affectionately, as pets.

There is an ironic contrast, here, between the affluent American servicemen pampering these birds just a few miles from the place (Pearl Harbour) which witnessed Japan's entry into the war; and the starving Japanese troops on Wake Island turning to the rails there for food. But the Laysan Rail, too, proved to be a war casualty. In 1943 a Naval landing craft drifted ashore on Midway and rats got on to both Sand and Eastern Islands. This was the end for the rails – the last ones were seen on Sand on 15 November 1943 and on Eastern in June the next year.

Meanwhile, despite contrary claims, it seems that no-one had thought of re-introducing the birds to their original home of Laysan where, by 1924, the rabbits had been exterminated, the vegetation had re-established itself and both the native finch and teal had staged a comeback. By this cruel series of chance events, the Laysan Rail was lost for ever.

Jamaican Wood Rail or Uniform Rail
Aramides concolor concolor

Laysan Rail or Spotless Crake *Porzana Palmeri*

Hawaiian O-O *Moho nobilis* Extinct *c.* 1934

Oahu O-O *Moho apicalis* Extinct *c.* 1837

Molokai O-O *Moho bishopi*
Extinct *c.* 1904

Hawaiian 0–0 *Moho nobilis*
Extinct *c.* 1934

Birds in Paradise

THE HONEYEATERS AND HONEYCREEPERS

In the popular imagination, the 'South Sea' Islands of the Pacific since the great age of exploration have always evoked the image of an earthly paradise. Of all these 'island paradises', none is more celebrated than the mid-Pacific group that Captain Cook named the Sandwich Islands, but now are known as the Hawaiian Islands.

The most striking aspect of the Hawaiian paradise was, and in some ways still is, the astonishingly unique and varied fauna and flora. The weathering lava and (in the western attols) coral limestone, acted upon by the subtropical temperatures and phenomenal rainfall, had given rise to dense and varied rain-forest, parkland and ravine areas wherein no less than 97 per cent of plant and tree species were endemic to the islands.

Threequarters of this natural forest was already lost by the 1950s to cultivation, cattle-browsing and fire; and if the islands lose much more of it they will lose their watersheds too, and rapidly erode into a wasteland. Already at least four tree species are extinct or survive as unique specimens. Each of them (Pritchard Palms and Hibisca-delphi) was vitally bound up – probably symbiotically – with the lives

Molokai O-O *Moho bishopi* Extinct *c.* 1904

of extinct birds. A minimum of 270 Hawaiian plant species, subspecies or varieties are extinct, and another 800 are endangered.

In historical times 68 native species of land birds lived on the islands: all seem to have evolved from only 15 ancestral colonizers within the last three million years (a fairly short time in evolutionary terms). With many we see exemplified the extreme ways in which birds in isolated territories may diverge from a common ancestor into specialized forms which utilize every possible ecological niche in the new habitat – what Darwin called 'adaptive radiation'.

The most fragilely beautiful Hawaiian birds were without doubt the Honeyeaters (Meliphagidae) and the Honeycreepers (Drepanididae). And it was these birds, above all others, that suffered most devastatingly from the consequences of European colonization.

KIOEA

Meliphagidae
Chaetoptila angustipluma
EXTINCT *c.* 1850
Island of Hawaii

The Hawaiian Honeyeaters were all members of the large Australasian family, *Meliphagidae*. The islands supported 2 genera and 5 species of this family but only one species survives today.

Of all the honeyeaters, the largest was the green Kioea (*Chaetoptila angustipluma*). It was also among the first birds to disappear from the islands after Europeans arrived. Very little is known about the Kioea, and only four specimens are ever known to have been collected.

It was confined to the island of Hawaii and, probably, to the restricted open plateau area between the forests and the true volcanic peaks. The only well-documented specimen was taken in 1840 by a naturalist called Peale, attached to the US exploring expedition. He wrote of it in 1848: 'This rare species was obtained in the Island of Hawaii. It is very active and graceful in its motions, and is disposed to be musical, having most of the habits of a Meliphaga. They are generally found about those trees which are in flower.'

The name 'Kioea' seems to stem from its use in the *Hawaiian Annual* for 1879 by a Mr Dole, who claimed that the bird was also found on Molokai. Other evidence suggests this is unlikely, and when we consider that Kioea was the name given by natives there to a migrant curlew and means 'standing high on long legs', it is not at all unlikely that this honeyeater has been saddled with a false name as well as with extinction.

O-OS

Meliphagidae
Moho nobilis
EXTINCT *c.* 1934
Hawaii
Moho bishopi
EXTINCT *c.* 1904
Molokai
Moho apicalis
EXTINCT *c.* 1837
Oahu

The other honeyeaters of Hawaii seem to have suffered from misnaming too: to this day the Latin name for the genus is 'Moho', which was the Polynesian name for the rails, whereas their true vernacular name was 'O-O'.

There were at least four O-Os – each restricted to one island. There were O-Os on Oahu (*Moho apicalis*); Molokai (*Moho bishopi*); and Hawaii (*Moho nobilis*). These three are now extinct. An unnamed species, also extinct, inhabited Maui – H. W. Henshaw saw one there around 1900. Of the entire genus, only the Kauai species (a dwarf form, *Moho braccatus*) survives: with two known living specimens. All honeyeaters were restricted to the mountain forests of their respective islands. The Oahu bird has not been seen since 1837; the Molokai since 1904; and the Hawaiian since 1934, though in each case there were unconfirmed reports for some years afterwards.

The causes of extinction are varied and can be analysed as similar to the honeycreepers', but we have direct evidence for the affect on the O-Os of introduced cattle, cats and rats.

The cattle seem to have altered the birds' habitat in some subtle way that we cannot pinpoint, by eating off the undergrowth (on which the tree-feeding O-Os were *not* dependent). R. C. L. Perkins wrote of seeing them in great numbers in 1892 in Hawaii: 'making with hosts of scarlet Iiwi, the crimson Agapane and other birds a picture never to be forgotten.' But a few years later he returned, at the same season, and wrote 'Although the trees were, as before, one mass of flower, hardly a single O-O was to be seen. The only noticeable change was that cattle were wandering over the flow and beginning to destroy the brushwood, just as they had already reduced the formerly dense forest bordering the flow to the condition of open parkland.'

At the same time Perkins went to Lanai and, in one ravine, found 22 native birds 'all killed by cats in the space of two days'. He shot two cats in the act of devouring O-Os and saw several others.

As for Molokai, scientists there found black rats actually moving about in the trees by daylight, a circumstance which speaks for itself.

But there is another factor to consider in any account of the O-O extinction, and this involves the native Polynesian Hawaiians and their centuries-old hunting of the birds. For the O-Os, especially the three extinct forms, were spectacularly beautiful – either brown or jet black and with long ornamental yellow plumes which the Hawaiians prized greatly. G. C. Munro saw a *Kahili* feathered stick in Kauai in 1891 made entirely from the central quills of the Hawaiian O-O, and the number of birds required to 'donate' enough feathers to make a traditional chieftain's cape was staggering; even though such garments might take years to make.

Moreover, the birds were easily caught – they would come to imitations of their distinctive calls; they would be kept alive for some time in captivity, fed on sugar cane juice, and used as decoys; and they were easily trapped with bird-lime on their favoured Ohia trees and Lobelia bushes.

The name O-O came from their call, described as 'owow, owow, ow!', the final note being a shriek which could be heard half a mile away. When feeding in large numbers they called almost continuously, especially in overcast weather, and were wonderfully agile and swift as they took nectar from the flowers on the wing. Quite often their heads were gauded with the pollen they had been feeding on, adding to their spectacular appearance.

The Hawaiian Honeycreepers are a family entirely endemic to the Hawaiian Islands. They are also considered the youngest bird family and demonstrate 'adaptive' and 'allotropic' radiation better than any family in the world. It is believed that there was only one ancestor, an American Honeycreeper, itself a product of adaptive radiation. It is therefore astonishing that the Hawaiian Honeycreepers should, 100 years ago, have comprised 9 genera, 22 species, and 64 subspecies. (Unfortunately, 3 genera, 9 species and 18 subspecies are now extinct.) Varying in size from 10 to 21.5cm (4–8½in.), these extremely beautiful and often brilliant birds demonstrate extreme diversity in structure – especially in bill shapes which have adapted to the specific habitats.

The foremost export on the honeycreepers, Dean Amadon, wrote:

'The Drepanididae are a rapidly evolving group living in a favourable yet dynamic and at times cataclysmic environment. . . . . When adverse factors (deforestation, predators and particularly disease) appeared with the advent of Europeans, declining species quickly became very rare or extinct; and others survived with varying success.'

As so often with island birds, the specialization that allowed them to survive in such large numbers and various forms in a limited space, proved a weakness when new forces or competitors intruded. The result was that 40 per cent of the honeycreeper species have become extinct, and a further 40 per cent are today in immediate danger of the same fate. Honeycreepers with the widest range, and without significant regional variations, have survived best.

As for the pressures towards extinction, the first and most obvious is destruction of the native forests. Less than a quarter of the original Hawaiian forests survive as the lowlands have given way to cultivation, and the mountains play host to goats, cattle, pigs and deer. The honeycreepers very often disappeared as soon as cattle entered an area of woodland; and the Koa forests, especially, are quickly destroyed by grazing. It was these forests that provided the major food source of the 'Koa finches' and many other honeycreepers.

Given this habitat destruction, the birds were even more vulnerable to competition, and since introduced birds (the 'Peking Robin' for example) swarm across the islands now, this may be a factor. However, introduced birds played a more indirect but sinister role in the decline of the honeycreepers, by carrying parasites and disease.

Several parasites have come with imported birds as well as the chicken-borne 'Bumblefoot' or Bird Pox. This disease is widespread and honeycreepers have been observed with its symptoms, but most writers now accept that the real killer has been avian malaria.

Before 1826 there were no mosquitoes on these islands. In that year the Night Mosquito came as a ship's stowaway to Maui and spread to the other islands. Avian (unlike Human) malaria can be spread by many genera of mosquitoes and the Night Mosquito is one of them. In 1938 imported pigeons were found to have the disease and there seems little doubt that the mosquitoes did the rest.

The only surviving Drepanididae live above the 600m (2000ft) limit of the mosquitoes' range and captured birds brought below that altitude are said to die invariably 'within days, of one of the insect-borne diseases'. It seems improbable, to say the least, that native Hawaiian birds could ever have the time to develop any strain of immunity to avian malaria.

Numerous other diseases seem to have swept the islands even before avian malaria spread, and many honeycreepers seemed notably susceptible to them, and to lack any natural immunities. G. C. Munro graphically describes how the birds of Kauai became exposed as early as the 1890s to introduced disease by coming to the forest edges and to low elevations: 'One was so disabled with lumps on legs and bill that it could scarcely fly. Another had a tumour a quarter of an inch thick in its throat, full of small worms, and a tumour on its ovaries containing a brown paste. Perkins found them with tapeworms.'

Lanai Alauwahio
Loxops (Paroremoyza) maculata montana
Extinct *c.* 1937
Molokai Alauwahio
Loxops (Paroreomyza) maculata flammea
Extinct *c.* 1970
Great Amakihi
Loxops (Viridonia) sagittirostris
Extinct *c.* 1900
Oahu Akepa
Loxops coccinea rufa
Extinct *c.* 1900

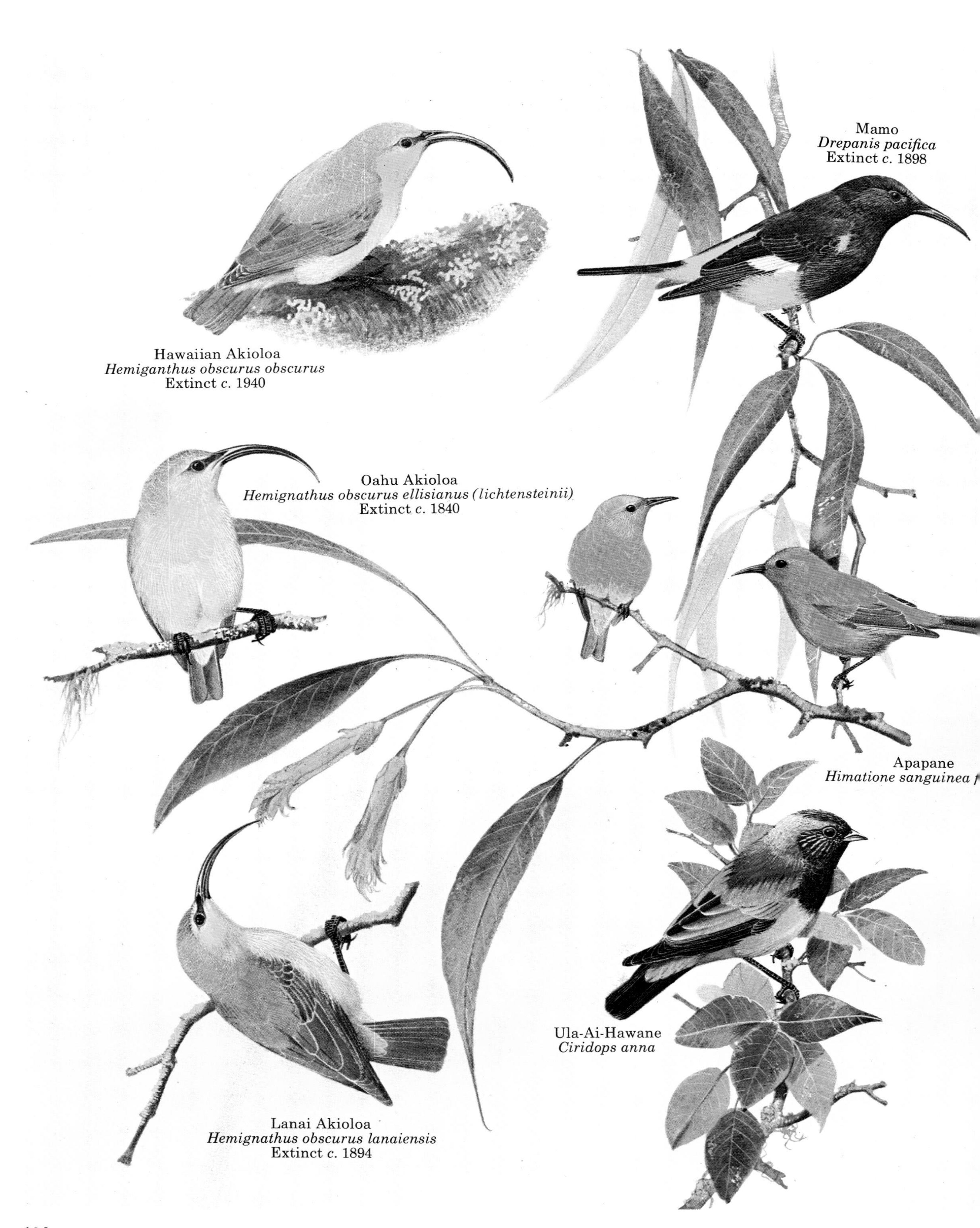
Mamo
Drepanis pacifica
Extinct c. 1898
Hawaiian Akioloa
Hemiganthus obscurus obscurus
Extinct c. 1940
Oahu Akioloa
Hemignathus obscurus ellisianus (lichtensteinii)
Extinct c. 1840
Apapane
Himatione sanguinea
Ula-Ai-Hawane
Ciridops anna
Lanai Akioloa
Hemignathus obscurus lanaiensis
Extinct c. 1894

Kauai Akioloa
Hemignathus obscurus procerus
Extinct *c.* 1965

Oahu Nukupuu
Hemignathus lucidus lucidus
Extinct *c.* 1890

Maui Nukupuu
Hemignathus lucidus affinis
Extinct *c.* 1896

Kauai Nukupuu
Hemignathus lucidus hanapepe
Extinct *c.* 1965

One additional factor must be considered, of course: namely direct persecution by human beings. While it is true that some honeycreepers were valued for their feathers; and that – as their scarcity became apparent – others were much sought after by collectors and their agents; these pressures seem to have been of relatively small significance to the birds' disappearances. The one notable exception was the case of the spectacular Mamo (*Drepanis pacifica*) which had always been pursued for its black and yellow plumage, the fabric of royal Hawaiian cloaks. It was rapidly hunted into extinction after the introduction of European firearms.

GREAT AMAKIHI

Drepanididae (Psittirostrinae)
Loxops (Viridonia) sagittirostris
EXTINCT *c.* 1900
Hawaii

ALAUWAHIOS

Drepanididae (Psittirostrinae)
Loxops (Paroreomyza) maculata flammea
EXTINCT *c.* 1970
Molokai
Loxops (Paroreomyza) maculata montana
EXTINCT *c.* 1937
Lanai

OAHU AKEPA

Drepanididae (Psittirostrinae)
Loxops coccinea rufa
EXTINCT *c.* 1900
Oahu

The Honeycreeper (Drepanididae) Family has a complex structure. Its classifications are perhaps most easily explained in terms of its two sub-family divisions: the *Psittirostrinae* (insect and seed-eaters) and the *Drepaniinae* (nectar feeders). The ancestors of the nectar feeders seem to have been the first group to colonize Hawaii, wind-borne from America, probably in the Upper Pliocene period. The other sub-family represents a second, later wave of invasion and had, presumably, to adapt to new styles of feeding to utilize the available habitat and food-sources. Over the millenia, enough species evolved to occupy every conceivable ecological niche on the islands. Yet the period of time this took is short, in evolutionary terms, and it should be emphasized that the sub-family distinctions are not absolute. Despite their spectacular adaptations of bills and behaviour, some Psittirostrinae still feed partly upon nectar; while a few of the Drepaniinae supplement their diets with insects.

The Psittirostrinae's insect or seed-eating habits are most obviously shown by the birds' bill structures. Other characteristics include dense and fluffy plumage and considerable sexual dimorphism – the males are generally brightly coloured yellow, orange or red and the females are olive-green.

The extinct members of this subfamily demonstrate 'allotropic radiation' in many aspects, but particularly in bill structure. The 19cm (7½in.) Great Amakihi or Green Solitaire (*Loxops sagittirostris* – extinct 1900) of Hawaii; and the smaller (10cm, 4in.) scarlet Molokai Alauwahio (*Loxops maculata flammea* – extinct 1970) and grey-green Lanai Alauwahio (*L. m. montana* – extinct 1937); were all bark-creeping birds with straight, sharp bills. ('Alauwahio' means 'creeper' in the Hawaiian language.) The 11.3cm (4½in.), orange-brown Oahu Akepa (*Loxops coccinea rufa* – extinct 1900) was, by contrast, slightly cross-billed.

AKIOLOAS

Drepanididae (Psittirostrinae)
Hemignathus obscurus obscurus
EXTINCT *c.* 1940
Hawaii
Hemignathus obscurus lanaiensis
EXTINCT *c.* 1894
Lanai
Hemignathus obscurus ellisianus (lichtensteinii)
EXTINCT *c.* 1840
Oahu
Hemingnathus obscurus procerus
EXTINCT *c.* 1965
Kauai

All four extinct subspecies of the 16.5cm (6½in.) Akioloa (*Hemignathus obscurus*) displayed extraordinarily downcurved bills, one third of the birds' total lengths, and both their Latin and Polynesian names mean 'curvebill'. The Hawaiian form (*H. o. obscurus*) was widespread in the island's mountain forests until about 1900, but has never been seen since 1940. The Lanaian Akioloa (*H. o. lanaiensis*) was last reported in 1894. There have been rumoured sightings in this century of the Akioloa of Oahu (*H. o. ellisianus*) but the last reliable report was in 1837. The most recent extinction from this genus has been the Kauai Akioloa's (*H. o. procerus*), for that bird seems to have succumbed to mosquito-borne disease as late as 1965.

Lesser Koa Finch
Psittirostra flaviceps
Extinct *c.* 1891

Greater Koa Finch
Psittirostra palmeri
Extinct *c.* 1896

Kona Finch
Psittirostra kona
Extinct *c.* 1894

NUKUPUUS

Drepanididae (Psittirostrinae)
Hemignathus lucidus lucidus
EXTINCT *c.* 1890
Oahu
Hemignathus lucidus affinis
EXTINCT *c.* 1896
Maui
Hemignathus lucidus hanapepe
EXTINCT *c.* 1965
Kauai

The three extinct subspecies of the Nukupuu (*Hemignathus lucidus*) also had long, decurved bills but differed from the Akioloas in that their upper and lower mandibles were of different lengths – the lower ones being uncurved, short and chizel edged. This development suggests a movement away from a purely nectar-reliant diet, and the one surviving Nukupuu (*H. l. wilsoni*) may owe its continued existence on Hawaii to its total transference to an insect diet.

The Oahu Nukupuu (*H. l. lucidus*) was the first to disappear, vanishing from the lower Koa forests around 1890. Shortly afterwards the yellow-headed Maui Nukupuu (*H. l. affinis*) went too – it has not been since 1896 although, unlike Oahu, Maui still preserves some virgin forest. The Nukupuu of Kauai (*H. l. hanapepe*) held out longer, though it was already rare by 1900. This, the brightest-coloured of all the Nukupuus, died out about 1965.

HAWAIIAN 'FINCHES'

Drepanididae (Psittirostrinae)
GREATER KOA FINCH
Psittirostra palmeri
EXTINCT *c.* 1896
LESSER KOA FINCH
Psittirostra flaviceps
EXTINCT *c.* 1891
KONA FINCH
Psittirostra kona
EXTINCT *c.* 1894
Hawaii

Three extinct species of this sub-family were specifically seed-eating birds which had developed powerful grosbeak or parrot-like bills. Inhabitants of the western Koa forests along Hawaii's Kona Coast, they were among the earliest Honeycreeper casualties, and are usually referred to as the 'Finches'.

The red-orange Greater Koa Finch (*Psittirostris palmeri*) was the largest bird in the whole family, at 21.5cm (8½in.), and was never seen after 1896. Its massive beak was used primarily to tear open the seed-pods on the Koa trees, and to break up the hard green beans or carobs inside. During the breeding season, however, it also took various caterpillars on and around the trees, and we may presume that these formed the main diet of the nestlings before they were equipped to deal with the tougher, vegetable substances. This Finch had a distinctive, flute-like whistling call which could be heard over a considerable distance.

The Lesser Koa Finch (*P. flaviceps*) was really a 'dwarf' version of its larger relative and its habits and diet seem to have been identical. It disappeared about five years earlier, not having been sighted since October 1891.

The smallest of this group was the Kona Finch or Chloridops (*P. kona*) yet it had the heaviest beak of them all. For, although it shared the other Finches' habitat, it did not feed as much on the Koa acacia trees, but mostly on the seeds of the Naio tree, sometimes foraging up to 1350m (4500ft) altitude to obtain them. These seeds are hard and dry and it requires a beak of nutcracker strength to break their casings. The Kona Finch also disappeared in the last decade of the century, and was last seen in 1894.

Apapane
Himatione sanguinea freethii

APAPANE

Drepanididae (Drepaniinae)
Himatione sanguinea freethii
EXTINCT *c.* 1925
Laysan

The more specifically nectar-reliant half of the Honeycreeper Family, the sub-family Drepaniinae, is notable for its often spectacular lanceolate and stiffened feathers which produce a characteristic 'whirring' sound in flight. These birds' bills are generally about the same length as their heads, and are distinctly decurved for reaching and extracting pollen. There have been four known extinctions from this sub-family.

The Apapanes (*Himatione sanguineae*) are most noted for the whirring sound of their wings. On most islands they are still fairly widespread, but the Apapane or 'Redbird' or Laysan (*H. s. freethi*), a much paler form than the other subspecies, lost its entire habitat through the browsing of introduced rabbits. In 1923 Alexander Wetmore could only find three individuals, and a sandstorm shortly thereafter seems to have completed the Redbird's extinction.

The Ula-Ai-Hawane (*Ciridops anna*), a small red Honeycreeper on Hawaii itself, probably disappeared about 1892. It was a monotypic species which depended on the Hawane palm-trees for its fruit-juice diet. Interestingly, several species of these trees are themselves on the brink of extinction.

ULA-AI-HAWANE
Drepanididae (Drepaniinae)
Ciridops anna
EXTINCT *c.* 1892
Hawaii

The Mamo of Hawaii (*Drepanis pacifica*) has already been mentioned as a victim of feather hunters and introduced firearms. One hunter with a shotgun is known to have shot in one day as many as 12 of these rare birds. The last definite sighting of the Mamo was in the Kaumau forest in 1898. Its cousin on Molokai, the Black Mamo (*Drepanis funerea*), whose beak was even longer and more decurved, was also hounded by collectors in its last days, and has not been seen since 1907.

Hawaii may still constitute a paradise in the eyes of package-tour operators, plantation owners and surfers, but the Koa forests are almost all gone and those which remain will never again witness the scene once observed by Dean Amadon when the Honeycreepers, 'high' on nectar, swarmed from treetop to ground level, vivid with different colours, 'constantly flittering' in excitement, calling, singing and 'whirring', like so many bright and fragile butterflies around a flowering tree in Eden.

MAMOS
Drepanididae (Drepaniinae)
MAMO
Drepanis pacifica
EXTINCT *c.* 1898
Hawaii
BLACK MAMO
Drepanis funerea
EXTINCT *c.* 1907
Molokai

Mamo
Drepanis funerea

Ula-Ai-Hawane *Ciridops anna*

Oahu Omao
Phaeornis obscurus oahensis
Extinct *c.* 1825
Lanai Omao
Phaeornis obscurus lanaiensis
Extinct *c.* 1931
Molokai Omao
Phaeornis obscurus rutha
Extinct *c.* 1936

The Singers

THRUSHES, WARBLERS, FINCHES AND ALLIED SONGBIRDS

If the honeyeaters and honeycreepers graced the forests of the Hawaiian paradise with their variety and colour, the Omaos or Hawaiian Thrushes added the dimension of sound. Forming an endemic genus (perhaps descended from the American *Myadastes*), these unimposing brown and grey birds sang year round among the Koa groves; and their music was as renowned for its variety as its beauty.

One witness described the 'series of flute-like notes . . . sometimes more like that of the Chinese thrush, although shorter and repeated'. They also had a 'whisper song' from the underbrush, and a loud, clear song from exposed perches – especially in December and May. The Omaos were most noted, though, for singing on the wing – an unusual habit in thrushes. They were said to 'quiver' their wings both when perched and when, in their flight-song, they arched up from 15m (50ft) perches and dropped, still singing, into the forest.

Each main island had its own distinct subspecies of omao, and Kauai had two. Three are extinct, the Kauai birds are all but gone, and only the omao of Hawaii itself has any real prospect of survival.

THE OMAOS

Turdidae

Phaeornis obscurus lanaiensis
EXTINCT *c.* 1931
Lanai

Phaeornis obscurus rutha
EXTINCT *c.* 1936
Molokai

Phaeornis obscurus oahensis
EXTINCT *c.* 1825
Oahu

In 1944, the naturalist G. C. Munro wrote of the Lanai Omao (*Phaeornis obscurus lanaiensis*): 'From 1911 to 1923 this bird was under my observation as I frequently rode the bridle trails to the forest. It was at that time a common bird and its call notes could be heard constantly . . . It declined in 1923 when the population of Lanai was increased and the town of Lanai City was built . . . I watched its decline till 1931. The few times I have been through the Lanai forest since 1931 the thrushes' call notes have been conspicuously absent.'

Indeed, 1931 was the last time anyone saw or heard the Lanai Omao. It was 17.8cm (7in.) long and its underside was a paler grey than its relatives'. Its feet and legs were blackish-grey. The Molokai Omao (*Phaeornis obscurus rutha*) may have survived as late as 1963, despite the habitat devastation on that island. It was one of the largest omaos, at 20.3cm (8in.), its throat and breast were a deeper grey than the Lanai bird's and its back was darker. Its feet and legs were pink. The Oahu Omao (*Phaeornis obscurus oahensis*) was the first Hawaiian Thrush to become extinct: it has not been seen or collected since 1825.

Like the honeycreepers, the omaos suffered drastically from exotic avian diseases spread by Night Mosquitoes; but, with their diet of beetles and fruit, they were affected too by forest destruction, especially on Molokai where early introductions of cattle and goats were followed, in 1868, by the release of Axis Deer. What the cultivators spared, these animals devastated. In 1920 the Deer were brought to Lanai too, evidently on the advice of G. C. Munro who later admitted it was 'the greatest mistake I ever made!'

LAYSAN MILLERBIRD

Sylviidae
Acrocephalus familiaris familiaris
EXTINCT *c.* 1920
Laysan, Hawaii

Molokai has been called 'a biological disaster area', but the outer Hawaiian island of Laysan, where rabbits were introduced in 1913, fared even worse. In ten years the rabbits had reduced Laysan to the 'barren waste of sand' described by Alexander Wetmore in 1923, and had caused the extinction of the chief songbird, the Laysan Millerbird (*Acrocephalus familiaris familiaris*) along with the endemic rail and honeycreeper. Wetmore, an ornithologist, found only 'two coconut palms, a stunted tree and an ironwood or two' and recalled: 'The desolation of the scene was so depressing that unconsciously we talked in undertones. From all appearances Laysan might have been some desert with the gleaming lake below merely a mirage.'

The millerbirds belong to the family of Old World Warblers (Sylviidae) noted, as their name suggests, for their melodic, liquid songs. The Laysan Millerbird was about 14cm ($5\frac{1}{2}$in.) in length, grey above and buffy white below, with a white superciliary stripe. It was totally insectivorous and got its name from its fondness for 'millers' – a kind of moth. The millerbird nested in grass tussocks.

Alfred M. Bailey was on Laysan early in 1913 and found the millerbirds still abundant. He wrote that they were: 'regular visitors to our table at mealtimes and to our workshops throughout the day, and so tame that if we remained quiet they would land upon our head. They searched in crannies for millers and caterpillars, their favourite food, and we often saw them over the portalaca flats bordering the lagoon. They were always extremely busy. It is probable that we were the last to see the species in life . . .'

Lord Howe Island and its satellites lie roughly equidistant from Australia and New Zealand in the Tasman Sea. Like many of the Pacific Islands, Lord Howe supported only bird life, and European contact was also its first human contact. The initial impact was considerable. By 1900, the larger endemic birds – the gallinule, pigeon and parakeet – had all been exterminated. By contrast, the island's four songbirds managed to survive predation by Man, dog, cat, goat and pig until 1918. In June that year, the SS *Mokambo* ran aground at Ned's Beach and was not refloated for nine days. During this period the rats got ashore, and within six years entirely exterminated the island's songbirds.

LORD HOWE ISLAND VINOUS-TINTED BLACKBIRD
Turdidae
Turdus (Poliocephalus) xanthopus vinitinctus
EXTINCT *c.* 1920
Lord Howe Island, Tasman Sea

One was the 22.9cm (9in.) thrush, the Lord Howe Island Vinous-tinted Blackbird (*Turdus* [*poliocephalus*] *xanthopus vinitinctus*) – whose name relates to the lavender ringe on its chestnut underparts. A. Basset Hull was on the island in October 1907 when he called this 'a very common and exceedingly tame species'. The islanders called it the 'Doctor Bird' and it was, Hull says: 'seen everywhere scratching among the dead leaves with the industry of a barnyard fowl, or perched on the low shrubs, emitting its sharp whistling chirp. Its local name is said to be derived from a sharp double knocking sound when the bird is alarmed. Though so tame that one can approach quietly within a few feet, any sudden appearance of a human being, or a loud noise, will send it scolding away for a short flight, but it soon stops to reconnoitre, curiously watching the intruder. It is also very suspicious, and will desert a nest either when building, with eggs, or even with young birds if touched by a human hand.'

Six years later Roy Bell made extensive (unpublished) field notes on his two year stay on the island. He noted in 1913 that the Doctor Bird or 'Ouzel' was common but not in great numbers. But when the rats got off *Mokambo* in 1918, the low-nesting blackbird was doomed and did not last more than two years.

LORD HOWE ISLAND FLYCATCHER
Acanthizidae
Gerygone igata insularis
EXTINCT *c.* 1920
Lord Howe Island

A. B. Hull also wrote about the other three songbird victims of the rats. One of these, the Lord Howe Island Flycatcher or 'Rainbird' (*Gerygone igata insularis*), was a 12.7cm (5in.) warbler which Hull separated into two subspecies (*insularis* and *thorpei* – the latter being smaller and having a yellow breast). Assuming he was right (though later writers seem to have ignored his observations) we are actually dealing here with two, not one, extinctions.

In any case the islanders did not distinguish between the two forms, calling them both Rainbirds or, in imitation of their songs, 'Pop Goes The Weasel'. Hull himself wrote of the 'sweetly mournful note' and called the song 'Chromatic but . . . staccato and varied in its tones.

'The birds are fairly numerous, flying briskly about the tree-tops; and are especially active in the pursuit of small insects after a shower of rain.' He was told by the islanders that the Rainbird's nest was dome-shaped, with an entrance at the side having a projecting hood.

LORD HOWE ISLAND WHITE EYE
Zosteropidae
Zosterops strenua
EXTINCT *c.* 1923
Lord Howe Island

Hull mentioned great numbers of the endemic White Eye (*Zosterops strenua*) on the island, noting: '. . . its powerful song makes music all day long in the palm-glades and on the wooded hillsides'. Also known as the Robust Silvereye, this was a bird of the trees, mostly green, but with bright yellow throat and white belly. The yellow markings were what most marked it off from related species. Although it was only 7.6cm (3in.) long, it was known as the 'Big Grinnell' by the islanders, to distinguish it from the considerably smaller Little Grinnell or Grey-breasted Silvereye.

Hull examined several White Eye nests, in Kentia palm trees, describing them as 'large, loosely constructed and cupshaped, composed outwardly of palm fibre, woven with dried grasses and lined with finer material of the same kind'. Some nested in shrubs overgrown with vines – a site particularly vulnerable to rats.

LORD HOWE ISLAND FANTAIL
Muscicapidae
Rhipidura fuliginosa cervina
EXTINCT *c.* 1924
Lord Howe Island

Almost the first object of natural history interest that Hull saw on the island was a Lord Howe Island Fantail's (*Rhipidura fuliginosa cervina*) nest which had been built in a small tree, about 4.6m (15ft) from the ground, and then abandoned. He found the new nest 'in a small prickly shrub' with the female sitting on two eggs which she reluctantly left. The nest he described as 'wine glass shaped, with no "foot" but with a rudimentary "stem"'.

The birds were characterized by their pale throats and the light cinnamon-brown colour of their whole undersurface. They were 12.7cm (5in.) in length with long, round-ended tails, characteristically held like a fan. Hull wrote: 'This Fantail is very tame, and fond of frequenting the vicinity of dwellings, where it will often enter the kitchen and capture flies from the walls.'

Six years after the rats came ashore H. A. Payten wrote to the Chief Secretary's Office in Sydney (10 May 1924) that the Fantails were 'practically wiped out'. Four years later a thorough search was made of the island, and none of these birds were found.

KITTLITZ'S THRUSH
Turdidae
Zoothera (Aegithocichla) terrestris
EXTINCT *c.* 1828
Peel Island, Bonin Group, North Pacific

BONIN ISLAND GROSBEAK
Fringillidae
Chaunoproctus ferreirostris
EXTINCT *c.* 1828
Peel Island, Bonin Group, North Pacific

About 1760km (1,100 miles) south of Tokyo Bay is the 27 island group called the Bonin Islands. Like Lord Howe Island, they were almost solely inhabited by birds until Europeans arrived.

The first records we have are from 1827 when HMS *Blossom* landed on the main island (Peel Island). It is from this ship's journals that we first hear of the soon to be extinct Bonin Wood Pigeon and Bonin Night Heron. It was not until the following year, when the naturalist J. H. von Kittlitz landed on Peel, that the songbirds of the Bonins were noted: Kittlitz's Thrush (*Zoothera* [*Aegithocichla*] *terrestris*) and the Bonin Island Grosbeak (*Chaunoproctus ferreirostris*).

Kittlitz collected four thrushes and named them '*Turdus terrestris*'. He commented on the large numbers of pigs running wild, along with the rats and feral cats. Neither songbirds, nor indeed the wood-pigeons and the night herons, were seen or heard again.

We must assume that this medium-sized thrush was restricted to Peel Island. Whether or not it was actually flightless, it certainly stayed on or near the ground and both it and its nests were, hence, especially vulnerable.

In 1889 the collector, A. P. Holst, spent four summer months on the Bonins, and his mention of introduced species suggests vividly what had happened to the island. He saw not only rats, but 'good sized deer . . . wild goats, wild boars, wild cats'. Though he collected several of the species discovered by *Blossom* and by Kittlitz, this was mostly on other islands of the group. The species restricted to Peel Island had largely disappeared, the thrush among them.

Laysan Millerbird
Acrocephalus familiaris familiaris
Extinct *c.* 1920

Lord Howe Island Vinous-tinted Blackbird
Turdus (Poliocephalus) xanthopus vinitinctus
Extinct *c.* 1920

Lord Howe Island Fantail
Rhipidura fuliginosa cervina
Extinct *c.* 1924

Lord Howe Island White Eye
Zosterops strenua
Extinct *c.* 1923

Lord Howe Island Flycatcher
Gerygone igata insularis
Extinct *c.* 1920

Holst searched in vain, too, for the Bonin Island Grosbeak which Kittlitz had collected. Our sole knowledge of this large (17.8cm, 7in.) finch comes from Kittlitz's *Kupfertafeln* . . . of 1832–3: 'This bird lives on Bonin Shima, alone or in pairs, in the forest near the coast. It is not common but likes to hide, although of a phlegmatic nature and not shy. Usually it is seen running on the ground, only seldom high in the trees. Its call is a single soft, very pure, high piping note, given sometimes shorter, sometimes longer, sometimes singly or sometimes repeated. In its muscular crop and spacious gullet I found only small fruit and buds.'

The male was quite striking with head, cheeks, throat and upper breast a reddish orange colour, a pale belly and the rest of the plumage olive brown, dark striped on the back. The female was all brown, with yellowish brow, and a generally mottled appearance.

BAY THRUSH

Turdidae
Turdus ulietensis
EXTINCT *c.* 1780
Raiatea, Society Islands

Sometime in the last week of May 1774 a shore party from the ships of Captain James Cook's second expedition put in to the Leeward island of Raiatea, 320km (200 miles) west of Tahiti, to survey and collect specimens. Among the animals they brought back was a female, thrushlike bird (*Turdus ulietensis*) which was duly catalogued on 1 June and then meticulously painted by G. A. Forster, the younger of the two German naturalists attached to the expedition.

The new specimen was also described in detailed notes by J. Latham and by the elder Forster. The bird has not been seen since and the specimen itself has disappeared: it entered Sir Joseph Bank's collection and, when that was deeded to the British Museum, was allowed to decay with the rest of that priceless stock of rarities. All that is left now is the excellent plate by Forster and the two eye-witness annotations.

J. R. Forster named it 'Turdus badius' or 'Chestnut Thrush', which adequately described its basic colour; but its actual family remains open to speculation: it was in some anatomical respects untypical of the Turdidae (thrushes), to such a degree that R. B. Sharpe (co-editor of the great nineteenth century monograph *Turdidae*) tried to identify it with the Mysterious Starling (*Aplonis mavornata*). But Forster was a good naturalist and examined the bird minutely. While acknowledging that in some ways the Bay Thrush reminded him of both the Graculae and Tanagrae, he concluded: 'Its appearance, plumage, habits and the soft song like a pan-pipe – all resembling the thrushes – persuade me that I must place this bird among the thrushes.'

The Bay Thrush was the size of the European Song Thrush (22.6cm, 8.5in.), a general chestnut colour, with a rounded, darker tail and 'dusky black' legs. Forster's terse summary in his *Descriptiones animalium* . . . tells us all that is known of the bird's habitat: 'It lives in the island Oradea (Raiatea) among the thickets, mostly in enclosed valleys. The islanders' name for it is "Eboonae-bou-nou".'

TONGA TABU TAHITI FLYCATCHER

Muscicapidae
Pomarea nigra tabuensis
EXTINCT *c.* 1800
Tonga Tabu (Tongatupu), Central Pacific

On that same expedition, nearly a year earlier, Cook's ship landed on Tonga Tabu (Tongatupu). Tonga Tabu is the largest of the Tonga islands in area (621sq km, 240sq miles) but is only 18.3m (60ft) above sea level at its highest point. The only endemic landbird known to have lived there was a now extinct flycatcher (*Pomarea nigra*

Tonga Tabu Tahiti Flycatcher
Pomarea nigra tabuensis
Extinct *c.* 1800
Bay Thrush
Turdus ulietensis
Extinct *c.* 1780
Chatham Island Fernbird
Bowdleria rufescens
Extinct *c.* 1895
Kittlitz's Thrush
Zoothera (Aegithocichla) terrestris
Extinct *c.* 1828
Bonin Island Grosbeak
Chaunoproctus ferreirostris
Extinct *c.* 1828

tabuensis) which was observed and collected by J. R. Forster in 1773.

Forster only procured a female specimen and, having described it in detail in his *Descriptiones Animalium*, continues: 'after close consideration I believe this must be the female of *Muscicapa atra* [the Tahiti Flycatcher of the Society Islands] or another variety of that species found on the island of Saint Christma or Weitaho, a very sweet, soft singer.' Forster is certainly an authority we can trust, and it is accepted by ornithologists today that his bird was in fact a *Pomarea*, a subspecies of the pied Tahiti form.

The cause of extinction is not known – but Tonga Tabu was not, anyway, an ideal habitat for such a bird: it may well have been one of a small, isolated group of birds, easily diminished in numbers by any disturbance, direct or indirect.

CHATHAM ISLAND FERNBIRD

Sylviidae
Bowdleria rufescens
EXTINCT *c.* 1895
Chatham Island, New Zealand

East of New Zealand is the Chatham Group of islands where a distinct species of Old World Warbler was discovered by Charles Traill in 1868. This was the Chatham Island Fernbird (*Bowdleria rufescens*). On Mangare Island, Traill spotted a small bird in the grass and stunted vegetation and knocked it down with a stone. It was forwarded to the New Zealand ornithologist, W. L. Buller, who described it as a new species.

Our only eye-witness account of the fernbird comes from the collector, H. H. Travers, who found it, in 1871, quite common on the small, steep Managre Island but much reduced on Pitt: 'Its peculiar habit of hopping rapidly from one point of concealment to another renders it difficult to secure. It has a peculiar whistle, very like that which a man would use in order to attract the attention of another at some distance . . . It is solitary in its habits and appears to live exclusively on insects.'

W. R. B. Oliver summarized the birds' decline: 'Through burning and the introduction of cats on Pitt Island, and collectors on Mangare, the Chatham Island Fernbird became extinct about 25 years after its discovery.' The last specimen taken was shot by one of Lord Rothschild's collectors after a climb along the precipice of northern Mangare Island, and hopes that some birds might survive up there, unmolested by cats, have proved unfounded.

This fernbird was distinguished from other forms by the virtual absence of spots on its underparts, and by its bright rufous upper surface. It was also reported several times to have produced part-albinos. In 1896 another collector, W. Hawkins, wrote: 'The Fern-bird is extinct. I spent a fortnight on the island where they used to be, but never saw any sign whatever of them.'

São Thomé Grosbeak
Neospiza concolor

SÃO THOMÉ GROSBEAK

Fringillidae
Neospiza concolor
EXTINCT *c.* 1900
São Thomé Island, Gulf of Guinea

The equatorial island of São Thomé, in the Atlantic Gulf of Guinea, was given over by its Portuguese colonizers to coffee-growing. As a result, most of the forests up to about 1500m (5000ft) have yielded to plantations. With the forests went an endemic finch, the São Thomé Grosbeak (*Neospiza concolor*), which has not been seen since 1900 despite frequent searches. Only two specimens of the São Thomé Grosbeak have ever been collected, and virtually nothing is known of its habits. The dense bush at higher altitudes could possibly still harbour the stocky songster, but most authorities accept its extinction.

Finally, there were two American finches that became extinct by the turn of the century. They were the Saint Kitts Puerto Rican Bullfinch (*Loxigilla portoricensis grandis*) and the Guadalupe Rufous-sided Towhee (*Pipilo maculatus consobrinus*).

The bullfinch was a common and very large (20.3cm, 8in.) finch, locally known as the 'Mountain Blacksmith'. It was a black bird, with its crown, chin, throat, neck and upper chest and under tail-coverts rufous. It had a heavy bill and its song was described as 'loud cardinal-like singing of six or seven notes. Call-note a soft tseet'. Introduced African monkeys were supposedly responsible, through nest predation, for this bird's extinction.

SAINT KITTS PUERTO RICAN BULLFINCH

Emberizidae
Loxigilla portoricensis grandis
EXTINCT *c.* 1900
Puerto Rico

The Rufous-sided Towhee of Guadalupe (off the coast of Baja, California) had a distinctive song aberration from its relatives': 'a single quick note' which made at least one ornithologist think there was a bluebird somewhere in the wood. The male towhee at the break of dawn 'mounted his throne on the topmost branch of a cypress and . . . sounded his morning trill'.

Although the towhee was a largish finch (17.8cm, 7in), it was extremely well-camouflaged by its plumage, which was said to resemble the patterns of dead leaves. It was not entirely restricted to the island's cypress grove, as it sometimes foraged for seeds and insects on the plateau below Mount Augusta, but this mobility was no defence against the numerous goats and feral cats.

The towhee was last sighted in 1897. Like all the island's birds it was, by that time, extremely rare, and is numbered among the 40 per cent of Guadalupe birds destroyed by goats, cats and Man.

GUADALUPE RUFOUS-SIDED TOWHEE

Emberizidae
Pipilo (Erythropthalmus) maculatus consobrinus
EXTINCT *c.* 1900
Guadalupe Island, Mexican Pacific

Saint Kitts Puerto Rican Bullfinch
Loxigilla portoricensis grandis

Guadalupe Rufous-sided Towhee *Pipilo (Erythropthalmus) maculatus consobrinus*

Stephen Island Wren *Xenicus lyalli* Extinct *c.* 189

The Timid and the Brash

WRENS AND STARLINGS

STEPHEN ISLAND WREN

Acanthisittidae
Xenicus (Traversia) lyalli
EXTINCT *c.* 1894
Stephen Island, New Zealand

At the northern tip of New Zealand's South Island, two miles from the spectacularly mountainous D'Urville Island in Blind Bay, Cook Strait, is Stephen Island. Though only about 2.6sq km (1sq mile) in area it rises abruptly from the sea to 300m (1000ft) in elevation. This wooded island was the only habitat of a unique species of New Zealand wren, which was simultaneously discovered and exterminated in the year 1894.

The twentieth-century authority on New Zealand ornithology, W. R. B. Oliver wrote of the Stephen Island Wren (*Xenicus lyalli*): 'The history of this species, so far as human contact is concerned, begins and ends with the exploits of a domestic cat.' And in 1895, the pathetic story had appeared in the *Canterbury Press*:

'At a recent meeting of the Ornithologists' Club in London, the Hon. W. Rothschild, the well-known collector, described this veritable *rara avis*, specimens of which he had obtained from Mr Henry Travers of Wellington, who, we understand, got them from the lighthouse-keeper at Stephen Island, who in his turn is reported to have been indebted to his cat for this remarkable ornithological

"find". As to how many specimens Mr Travers, the lighthouse-keeper and the cat managed to secure between them we have no information, but there is very good reason to believe that the bird is no longer to be found on the island, and as it is not known to exist anywhere else, it has apparently become quite extinct. This is probably a record performance in the way of extermination. The English scientific world will hear almost simultaneously of its discovery and of its disappearance, before anything is known of its life-history or its habits . . . It was not a flightless bird, but from its structure it was evidently very weak-winged and thus fell an easy prey to the lighthouse-keeper's cat . . . all New-Zealanders who take an interest in the preservation of whatever is specially characteristic of the colony will deplore the extermination of such a creature. It is, indeed, saddening to reflect how one by one the rare and wonderful birds which have made New Zealand an object of supreme interest among scientists all over the world are gradually becoming extinct . . . And we certainly think it would be as well if the Marine Department, in sending lighthouse-keepers to isolated islands . . . were to see that they are not allowed to take any cats with them, even if mouse-traps have to be furnished at the cost of the State.'

A certain amount of debate later arose between the great collector, Walter Rothschild, and the New Zealand authority, Sir Walter Buller, over who first named the bird and over the number of extant specimens. In both cases Buller proved correct.

The discussion over numbers, in fact, raises certain suspicions about the collector, Mr Travers. Rothschild wrote: 'I have had a letter from Mr Travers, who tells me that all the 11 specimens (ten in my museum and one from which Sir Walter Buller made his description) were brought in by the cat. The bird itself has only been seen on two occasions alive, when disturbed from holes in the rocks, and was not obtained. It is nocturnal, runs like a mouse and very fast, and did *not* fly at all.' Buller commented on this: 'I think that Mr Rothschild is under a misapprehension in supposing he possesses all the known specimens except that described by me in the *Ibis*. Besides a pair in my son's collection, I purchased a specimen from Mr Henry Travers for Canon Tristram, and the former gentleman has since offered to sell me two more.'

Thus Travers handled at least 16, not 11 of the wrens. There is something rather suspicious concerning the acquisition of the last few specimens. As we have repeatedly seen, endangered animals have not infrequently been killed off by such collectors. Somewhat defensively Buller wrote of the extinction: 'I cannot see that any share of the blame attaches to Mr Travers . . . It is acknowledged that the offending cat would have devoured them all, if Mr Travers' agent had not been there to rescue some for science.'

Interestingly, in the public concern that briefly flared up in response to the Stephen Island Wren's extinction, an essay by a Mr Parnell was also published in the *Canterbury Press*:

'If the government would act as promptly in stopping marauders, commonly called natural history collectors, from visiting the outlying islands and carrying off the Tuataras and rare birds by hundreds, as it did the seal-poachers in the southern islands last year, it would gain the gratitude of science and coming generations.'

The 'Wrens' of New Zealand are not true wrens at all, but an eccentric family of birds (Acanthisittidae) that look like, and behave in a similar way to, the Common Wren. Only two of these 'wrens' now survive: the common Rifleman and the rare Rock Wren of Fiordland. The other forms, the Bush Wrens of North Island (*Xenicus longipes stokesi*) and Stewart Island (*X. l. variabilis*), have, with the introduction of cats, rats and stoats, suffered the same fate as the Stephen Island Wren.

The New Zealand Bush Wren was discovered on Captain Cook's second voyage by his chief naturalist J. R. Forster, while Cook refitted his ships in Dusky Sound. Forster's son. G. R. Forster, painted the bird and his plate was the basis for its being described as a 'Long-legged Warbler', whence its Latin name – *longipes*.

The North Island subspecies was first collected in 1850 in the Rimutaka Range. These two specimens were the only ones ever taken, despite numerous sightings. They were sent to the British Museum by Captain Stokes of the surveying ship *Acheron*.

The North Island Bush Wren differed from the Stewart Island subspecies in its 'shining slate blue sides of the neck and chest tinged, in certain lights, with greenish; and the patch of pure yellow feathers on the yellowish green flanks'.

The Stewart Island or Stead's Bush Wren seems to have survived on Stewart Island and its satellites until 1965. Some time before 1963 rats had been accidentally introduced, and are the probable cause of the extinction.

The New Zealand Bush Wrens, like all the Acanthisittidae, were primarily ground feeders and insectivorous. They had dome-shaped nests with concealed side-entrances in a variety of places (creepers, hollow logs, tunnels and burrows) but all were near the ground. They were also observed in small parties flying from tree to tree.

NEW ZEALAND BUSH WRENS

Acanthisittidae
Xenicus longipes stokesi
EXTINCT *c.* 1900
North Island, New Zealand
Xenicus longipes variabilis
EXTINCT *c.* 1965
Stewart Island and satellites

North Island Bush Wren
Xenicus longipes stokesi

The Guadalupe Wren (*Thryomanes bewickii brevicaudus*) was another bird which fell victim primarily to cats. However, even more than with the Stephen Island Wren, we may suspect that ornithologists, eager to obtain rare specimens, dealt the final blow.

When Guadalupe (off the Baja California coast) was first visited by Dr Palmer in 1875, two specimens were taken near the centre of the island. At that time the wrens were fairly widely distributed along the ridge-top undergrowth, though introduced goats were rapidly infringing on their territory. Ten years later, W. R. Bryant searched the whole island thoroughly and found wrens rare; the few survivors being confined to the northeast end of the island where pines straggled along the sharp ridge of North Head. Bryant took seven specimens, and 'Fearing the extermination of the species the balance of the colony was unmolested'! But the goats continued to encroach on the undergrowth, and the more deadly cats, which began to spread around 1875, came in. Loss of cover made the wrens liable, too, to be swept from the island by the frequent fierce gales.

At this point the Guadalupe Wren's extinction was probably a foregone conclusion. However, two men seem to have left nothing to chance. Near the end of May 1892, Clark P. Streator and A. W. Anthony went for one day to the North Head and hunted down the last three birds in the colony. Anthony afterwards visited the island

GUADALUPE (BEWICK'S) WREN

Troglodytidae
Thryomanes bewickii brevicaudus
EXTINCT *c.* 1892
Guadalupe, Mexican Pacific

Guadalupe (Bewick's) Wren
Thyromanes bewickii brevicaudus

several times and searched with assistants along the entire ridge. 'We never found any signs of the species which must now be classed among those that were' – and which he, in fact, had finished off.

The original observer, Dr Palmer wrote of their habits: 'They live in the brush, being rarely seen on trees . . . Their motions are very quick; their general habits restless and shy', and he noted their 'almost incessant activity'. Bryant called them 'timid rather than shy', and described their fleeing into the thickest brush as he scrambled over windfall pine trees, crushing the dry branches. 'When all was quiet they would cautiously approach until within a few feet of me, seemingly prompted by curiosity.' The only sound he heard from them was a few 'twit twits' from a startled female.

WEST INDIAN WRENS

Troglodytidae
Troglodytes aëdon (musculus) martinicensis
EXTINCT *c.* 1900
Martinique
Troglodytes aëdon (musculus) mesoleucus
EXTINCT *c.* 1971
St Lucia

In the West Indies, two Wrens have become extinct: the Martinique and St Lucia Wrens (*Troglodytes aëdon martinicensis*, and *T. a. mesoleucus*). These Wrens were once common birds on their respective islands. Again forest destruction, cats, rats and, in this case, mongooses were responsible for their extinction.

The last Martinique Wren was taken in 1886 when they were still numerous, but decline must have been rapid for by the end of the century they were gone. The Guadeloupe form (*T. a. guadeloupensis*) was long believed extinct, but its recent rediscovery has been offset by the disappearance, about 1971, of its St Lucian cousin.

RÉUNION CRESTED STARLING

Sturnidae
Fregilupus varius
EXTINCT *c.* 1868
Réunion

One of the most spectacular and unusual birds to disappear from the Mascarenes and one which survived until about 1870, was the Crested Starling (*Fregilupus varius*) of Réunion. It was almost as unlike a starling as the Dodos were unlike doves, being very nearly a foot long and characterized by an erectile crest of 'decomposed' white-shafted feathers which led the islanders to call it the Hoopoe, a bird which – to the layman – it very much resembles. Its distinctly decurved (down-curved) beak is equally unusual for a starling.

Nevertheless, close skeletal study has shown that this 'Hoopoe' was definitely of the starling tribe, most closely related to the Pastor group, but much altered by long-isolation and adaptation.

These birds were primarily ground feeders (though well able to fly) and were to be found in large flocks both in the forests of the uplands and mountains, and on swampy fields throughout the coastal plain. They were extremely trusting and easily killed with sticks and, because of their size, were not disdained by the islanders as a food source.

However, the 'Hoopoes' seemed to retain their numbers despite human persecution until the 1830s when they started to decline. This may have been partly to do with the dense cultivation of the plains, and the increase in human population, but since their numbers dwindled in the forest too, it seems likely that rats were their most destructive enemy.

The Crested Starling came close to being established on Mauritius where it might have survived: in 1837 J. Desjardins wrote from Mauritius: 'My friend, Marcelin Sauzier, sent me four alive from Bourbon [Réunion] in May 1835: They eat anything. Two escaped some months afterwards, and it might well happen that they will stock our forests.' But the birds were shot by local inhabitants

Réunion Crested Starling
Fregilupus varius
Extinct *c.* 1868
White Mascarene Starling
Necropsar leguati
Extinct *c.* 1880

the year that Desjardins wrote this.

In 1868 Pollen and Van Dam wrote in their *Fauna of Madagascar and Dependencies*: 'This species has become so rare that one has not heard them mentioned for a dozen years. It has been destroyed in all the littoral districts, and even in the mountains near the coast. Trustworthy persons, however, have assured us that they must still exist in the forests of the interior, near St Joseph. The old Creoles told me [Pollen] that, in their youth, these birds were still common, and that they were so stupid that one could kill them with sticks. They call this bird the "Hoopoe". It is, therefore, not wrong that a distinguished inhabitant of Réunion, Mr A. Legras, wrote about this bird with the following words: "The Hoopoe has become so rare that we have hardly seen a dozen in our wanderings to discover birds; we were even grieved to search for it in vain in our museum".'

A long-time resident of Réunion, M de Cortimoy, wrote in 1911: 'When I was a child, I knew this bird to inhabit the forest and live in flocks; their song was a clear note and they had a beautiful plumage. The bird was very tame and, being young, I killed dozens of them. I used to keep them in a cage without trouble. They eat bananas, potatoes, cabbages etc.'

We can only regret that more birds were not kept in captivity, and bred; for in the wild the last mention of the bird is that quoted by Pollen in 1868, and that may well have referred to several years earlier.

WHITE MASCARENE STARLING

Sturnidae
Necropsar leguati
EXTINCT *c.* 1840
Rodriguez

The anonymous author of the *Relation de l'Île de Rodrigue*, referred to several times before, described a bird which seemed to have vanished without trace by the start of the nineteenth century: 'A little bird is found which is not very common, for it is not found on the mainland. One sees it on the Islet au Mat, which is to the south of the main island, and I believe it keeps to that islet on account of the birds of prey which are on the mainland, as also to feed with more facility on the eggs, or some turtles dead of hunger, which they well know how to tear out of their shells. These birds are a little larger than a blackbird, and have white plumage, part of the wings and tail black, the beak yellow, as well as the feet, and make a wonderful warbling. I say a warbling, since they have many and altogether different notes. We brought up some with cooked meat, cut up very small, which they eat in preference to seeds.'

No other descriptions exist of this bird, which may well have gone into extinction along with the attol tortoises on which it seems to have been so dependent. From the description it could have been anything from a small oystercatcher to a diminutive chough or other crow.

Then in 1874 the Revd H. H. Slater, who was attached to a Transit of Venus expedition on mainland Rodriguez, dug up some subfossil bones (while recovering Solitaire bones from a cave) which proved to be those of a starling: a smaller relative of *Fregilupus*. Gunther and Newton wrote in the *Philosophical Transactions of 1879*: 'The discovery of an extinct Starling in Rodriguez, allied to the *Fregilupus* of Réunion, which appears to have held out a little longer in struggling for its existence within so narrow limits, is undoubtedly one of the most interesting results of Mr Slater's labours . . . *Necropsar* is altogether a smaller bird than *Fregilupus*, to which it is most closely allied.'

Four years earlier Newton had written of the Islet au Mat birds 'I am at a loss to conjecture what these birds were unless possibly some form allied to *Fregilupus*', and Slater's discovery went some way to supporting the idea that the mysterious Turtle Birds were starlings.

Then in 1897, H. O. Forbes, director of the Liverpool Museum, discovered an unknown, but well-preserved bird skin in the Derby Museum collection recently purchased from Lord Derby. Its label said that it had been bought in 1880 from the Parisian dealer M. J. Verreaux, and stated cryptically on the reverse 'Madagascar'.

Forbes discovered that Verreaux had visited the Mascarenes round about 1832, had collected birds there, including a *Fregilupus varius*, and tended (if he had not shot the bird personally) to label his specimens 'Madagascar' to indicate the general area.

Rothschild drew Forbes' attention to the 'Islet au Mat' passage and he wrote 'our Derby Museum skin so far agrees in its general white colour, in its yellow feet and bill, and in its size being near a Blackbird's; but it differs in having no black on the wings or tail.' In fact the colour of those parts of this specimen is a slight rusty shading. Forbes distinguished his specimen from the Bourbon Crested Starling by the form of the wings, the absence of the crest, the much less decurved bill and, of course, the size. It is only 22.9cm (9in.) long.

There remained problems of identification, however: this could not be the Turtle Bird, because of its colour; nor could it be the Rodriguez mainland starling of Slater because it was significantly smaller.

Many hypotheses have been put forward, some cautious, others wildly fanciful. We shall never know for sure. It is not impossible that *Necropsar leguati* was an aberrant colour form from Islet au Mat or one of the other islets; nor can the possibility be ruled out that it was an immature of *Necropsar rodericanus* (Slater's form) though the specimen does not seem to be a juvenile.

Perhaps the most likely possibility is that (as with the Bourbon Crested Starling) there was a strong sexual dimorphism and that the white starling is the female of its species.

On balance, then, *Necropsar leguati*, the White Mascarene Starling, was probably an inhabitant of mainland Rodriguez, perhaps less specialized in its dependence on tortoises than its black and white attol cousin, but beset by a far greater horde of introduced enemies. It is unlikely to have survived much beyond the 1830s when it was collected.

Still surrounded by mystery, explicable only by guesswork, the solitary skin in the Derby collection has understandably been proposed as 'the rarest bird in the world'.

KUSAIE MOUNTAIN STARLING

Sturnidae
Aplonis corvina
EXTINCT *c.* 1827
Kusaie, Caroline Islands, Southwest Pacific

The Caroline Islands in the southwest Pacific seem to have been reached by three waves of starlings. Two still survive, but the original colonizer, the glossy black Mountain Starling of Kusaie Island (*Aplonis corvina*) which is only known from the five specimens obtained by the German naturalist, von Kittlitz, who in 1827 gave it

Kusaie Mountain Starling
Aplonis corvina

its present name.

The bird has not been heard of since: Otto Finsch did not see it in his trip through Kusaie in March 1880 though he did not reach the interior mountains as Kittlitz had. W. F. Coultas of the Whitney Expedition did search the interior thoroughly in 1951, without success, and he reported enormous numbers of rats.

In 1890 Sharpe called it 'strictly a mountain bird', following Kittlitz's brief account, and this is probably to be explained by its being driven inland by the later arrival of the gregarious, aggressive though smaller Micronesian Starling (*A. opacus*). It shows no affinities with any other starlings of Micronesia, and is distinguished above all by its long bill. In this, as in size and colouring, it most resembles the Solomon's starling (*A. atrifuscus*), and its ancestors probably came down via the Solomons.

LORD HOWE ISLAND STARLING

Sturnidae
Aplonis fuscus hullianus
EXTINCT *c.* 1925
Lord Howe Island, Tasman Sea

Lord Howe Island Starling
Aplonis fuscus hullianus

The Lord Howe Island Starling (*Aplonis fuscus hullianus*) was a 17.8cm (7in.) bird, known to islanders as 'Cúdgimarúk', in imitation of its call. They also called the starlings 'Red-eyes'. A. B. Hull said of its appearance: 'Its soft, slatey-grey plumage, darker in the male than in the female, is somewhat at variance with its bright orange-red eyes; and its assertive manner, attitudes and loud challenging notes are not in keeping with its sober coat.'

He called it 'a bold and noisy marauder' and described it 'creating havoc among orchards and banana groves'. When disturbed or shooed-off the Cúdgimarúks would fly off to neighbouring trees and scold and return very soon. They lived, fed and moved about in pairs, especially during the nesting season.

Obviously the islanders would have wanted neither the motivation nor the opportunity to shoot these birds but there seems no doubt that it was rats, not human persecution that destroyed them. Not many years after the rats came ashore in 1918 there was not a single starling left and a clue to their vulnerability may lie in the fact that their nests were often close to the ground (simple, open structures in the 'hollow spout' of a dead limb) and that they always returned to the same nest and would lay again if their eggs were robbed.

Their bluish, red-blotched eggs were in clutches of four or five. The head and body of the bird much resembled the common starling's, but the wings were brown and the tail grey, tipped with brown.

MYSTERIOUS STARLING

Sturnidae
Aplonis mavornata
EXTINCT *c.* 1780
Society Islands (?)

Secret instructions handed by the Earl of Sandwich to Cook as he embarked on his third and last voyage (1778–9) included this: 'You are carefully to observe the nature of the soil and the produce thereof; the animals and fowls that inhabit or frequent it . . . and in case there are any peculiar to such places, to describe them as minutely, and to make as accurate drawings of them, as you can.' On his first voyage Cook had Banks and Solander as naturalists; on the second he had the two Forsters and Sparrman; but on this last voyage only the surgeon, William Anderson, had any kind of competence. J. R. Forster had probably taught him how to skin and preserve animals seen and collected. But he fell ill with consumption and died in the Bering Sea on August 1st 1778. When the *Resolution* and *Discovery* headed for the north Pacific from the Sandwich Isles early that year, they were entering an exotic territory only crossed before by the German naturalist, Georg Steller.

When the expedition returned in August 1780, their collection of birds and plants was quickly shared out between Sir Joseph Banks and Sir Ashton Lever (whose museum in Leicester Square was the finest collection in Britain and contained many birds from Cook's previous voyages). Both collections had a sad later history: the Lever collection was put up for auction when Sir Ashton got into financial difficulties and was dispersed. Banks' collection was given to the British Museum. According to R. B. Sharpe the specimens were probably 'inadequately prepared, always mounted, and, from a lack of proper appreciation of their priceless value, were allowed to decay, through a want of proper curatorial knowledge'.

It seems that all the specimens perished except one – *Aplonis mavornata*, a starling or starling-like bird, collected from an unknown Pacific island and never seen since. Its mystery is enhanced by the fact that Sharpe was convinced that this bird was a thrush, identical in fact to Latham's Bay Thrush (*Turdus ulietensis*). By this time, though, the specimen was in a sorry state. In Sharpe's words it 'has persisted in a kind of mummified state to the present day, after having been mounted and exposed to the light and dust of the old British Museum for nearly a century.'

But Sir Walter Buller before then, and E. Stresemann as recently as 1949, examined the specimen too and concluded that it was a starling; and Stresemann shows clear distinctions between the colour and size of this bird and the Bay Thrush as illustrated by Forster. In death as in life, then, this small brown starling retains its mystery. Where it came from and why it perished no amount of *post facto* detective work can now determine.

Mysterious Starling *Aplonis mavornata*

Heath Hen *Tympanuchus cupido cupido* Extinct *c.* 1932

The Game Birds

GROUSE, QUAILS AND DUCKS

Until the second half of the eighteenth century, Heath Hens and Prairie Chickens inhabited nearly all the open prairie east of the Rocky Mountains. They bred and fed in groupings several thousand strong, and the plain was alive in springtime with the characteristic uproar of hundreds of male birds 'booming' out their mating dance together, while strutting and displaying themselves to the females. Today, of the four distinct subspecies that made up the Heath Hen and Prairie Chicken super-species (*Tympanuchus*), three are endangered and still declining, and one – the Heath Hen (*Tympanuchus cupido cupido*) – has been extinct since 1932.

The Heath Hen inhabited most of the open, dry land of New England where much of the early settlement of America began, and were shot in their thousands by settlers, pioneers, and professional market hunters. The birds were 'so common on the ancient brushy site of Boston, that . . . servants stipulated with their employers not to have Heath Hen brought to the table oftener than a few times a week.' With the vast influx of European immigrants to eastern America, the Heath Hen population, however numerous, could not hold out

HEATH HEN

Tetraonidae
Tympanuchus cupido cupido
EXTINCT *c.* 1932
New England States, USA

long. As early as 1791 a few enlightened Long Islanders realized that something must be done if the Heath Hen were to last more than a few more decades. They attempted to introduce a closed season on the hunting of the bird – unfortunately the law had no effect whatever on the attitudes prevalent at the time. As J.C. Greenway noted on this event: 'A sharp focus is brought to bear on the state of mind of the people of those days when conservation was discussed. It is recorded that when the chairman read the name of the bill "An Act for the preservation of the heath-hen and other game", the northern members were astonished and could not see the propriety of preserving "Indians or any other heathen".'

Over-hunting apart, by 1800 many other forces threatened the Heath Hen's dwindling numbers. The Europeans brought with them their familiars: the dog, cat and rat; and the Heath Hen, being a ground-nester, proved to be especially vulnerable to these new predators – particularly the cat. There is evidence, too, of epidemic diseases among the native birds which very likely had their origin in introduced pheasants and chickens. But the ultimate factor was habitat destruction: the conversion of prairie into farmland.

By 1830, the Heath Hen was extinct on the mainland, but lived on the protected island of Martha's Vineyard in Massachusetts. Tim Halliday in his recent *Vanishing Birds* recounts the last years of the Heath Hen on this island: 'In 1890 there were estimated to be 200 birds on Martha's Vineyard; this fell to less than 100 by 1896. In 1908, by which time there were only about fifty birds left, a 1600-acre reservation was established on the island. This was a highly successful venture and by 1915 the Heath Hen population had risen to 2000. In 1916, however, naturally occurring climatic changes completely altered the situation. In the summer a bush fire devastated the heart of the breeding area and this was followed by a severe winter whose effect on the population was exacerbated by the arrival on the island of unprecedented numbers of Goshawks that took many of the hard-pressed Heath Hens. The population was reduced to less than 150 birds and, to make matters worse, a disease called Blackhead broke out, having probably been inadvertently introduced with turkeys. By 1927 only thirteen Heath Hens were left, by 1928 there were only two, and in 1930 there remained just one member of what had once been a spectacularly abundant species. It attracted widespread attention from ornithologists and tourists who gathered to see it and at least once it narrowly escaped the ignominious fate of being run over by a car. This last of the Heath Hens seems to have survived until 1932.'

In 1916, the noted American ornithologist Edward Forbush had quite reasonably predicted that the California Condor and the Whooping Crane would not survive more than two decades, and at the same time congratulated the Massachusetts authorities for saving the Heath Hen by their enlightened handling of this bird on Martha's Vineyard. By some miracle both the condor and the crane are still with us – though just as tenuously as they were in Forbush's time. And by a series of mischances, the Heath Hen is extinct instead.

As late as 1925, last ditch efforts were made on Martha's Vineyard to save the remaining Heath Hens. Special wardens were employed to control predators, and 120 cats were killed on the reserve.

Astonishingly, at such a critical time – there could not have been more than twenty Heath Hens left in the world by then – they discovered evidence of poachers working on the reservation. The protective measures were, of course, far too late: the inbred and sick population could not sustain itself. The last Heath Hen, an eight year old male, disappeared on the evening of 11 March 1932.

NEW ZEALAND QUAIL

Phasianidae
Coturnix novaezelandiae
EXTINCT *c.* 1868
New Zealand

In New Zealand, the only indigenous 'game bird' was the New Zealand Quail (*Coturnix novaezelandiae*) or 'Koreke' as the Maoris called it. It was found throughout the scrubland areas of New Zealand, but was especially abundant on the South Island. As one early colonist recalled: 'In the early days on the plains . . . a bag of twenty brace of Quail was not looked upon as extraordinary sport for a day's shooting.' In 1848, on the estate of Sir Edward Stafford near Nelson, a hunt was arranged '. . . and in the course of a few hours the party bagged 29½ brace'.

The first sighting records of the Koreke were made by Sir Joseph Banks in his journal of Cook's first expedition in 1769. Ninety-nine years later, the ornithologist Sir Walter Buller came suddenly on a covey near Blueskin Bay. He shot all the birds as specimens for his collection. It was the last time anyone saw the New Zealand Quail alive.

The Koreke was a dark quail about 20.3cm (8in.) long and 210–240g (7–8oz) in weight. It lived in open grass and scrubland, and in stormy weather hid among thick tussocks. Its call was heard most often in wet weather and was described as 'twit, twit, twit, twee-twit repeated several times in quick succession'.

New Zealand Quail *Coturnix novaezelandiae*

In 1871, the naturalist T. H. Potts wrote of the nearly extinct Koreke: 'On the ground their movements are active; sometimes they may be seen indulging in a dust bath as they lie basking in the sun; unless suddenly startled they almost always maintain that plump rounded appearance....

'The flight of the Quail is low, it used to be said that it would not rise after being flushed the third time: numbers were killed by sheep and cattle dogs in the early days, when it abounded.... We have seen it escape the talons of the Quail-hawk, by dropping perpendicularly, just when about to be struck, when all hope of escape from its relentless pursuer was quite abandoned.'

Had anyone then seen fit to arrange what today is called a 'captive breeding program', the species might well have survived, since the birds were not difficult to sustain in captivity. As Potts noted: 'In confinement, they are fond of picking about amongst sand, thrive well on soaked bread, grain of various kinds, and the larvae of insects;' though he added: 'the male is not an attentive mate at feeding time, and where several are kept in the same enclosure, constant little bickerings take place without actual hostilities being indulged in.'

The Quail's decline was extraordinarily rapid, for it was said to be still numerous in many places in 1860. Apart from hunting, fire (often accidental) is generally cited as the chief cause of the Koreke's extinction: 'Bush fires, extending often for many miles, must have been the active agent in destroying a bird possessing such limited powers of flight, as our handsome little quail.'

Frequently fires destroyed not only live Korekes and eggs, but the birds' food source and sheltering cover as well. There were other factors, too. The introduction of dogs, cats, rats and weasels had considerable impact on the Koreke's population. Then too, there is the probability of epidemics of one or more deadly avian diseases, since pheasants (and partridges) were introduced to New Zealand by sportsmen from 1842 onward. It is known that at least three diseases can be transferred from pheasants to quail with fatal results.

HIMALAYAN MOUNTAIN QUAIL

Phasianidae
Ophrysia superciliosa
EXTINCT *c.* 1870
Eastern Punjab, India

Strangely enough, the year 1868 marked the last sighting not only of the Koreke but also of the only other extinct member of the Quail family – the Himalayan Mountain Quail (*Ophrysia superciliosa*).

The Himalayan Mountain Quail was a dark game bird with a conspicuously red beak and red feet. It lived in the eastern Punjab in the foothills of the Himalayas. It was not, in fact, a true quail, but perhaps closer to what might be called a pigmy pheasant. It was 30.5cm (12in.) in length, with a relatively long tail, and was slightly smaller than a European Partridge. The male bird had distinctive head markings.

This Quail, the only species of its genus (*Ophrysia*), fed on grass seeds, berries and insects. It was seen in coveys of half a dozen in patches of long grass and brushwood on steep hillsides. It flew only reluctantly, when practically trampled upon – and then heavily and only for short distances. Its alarm note was 'a shrill whistle' and its contact note when feeding was 'low, short, and quail-like'.

Although some are known to have been kept in the Earl of Derby's private zoo in 1846, the Himalayan Mountain Quail was a little-

Himalayan Mountain Quail *Ophrysia superciliosa* Extinct *c.* 1870

observed species. Of a shy and withdrawing nature, it generally kept clear of men. Only twelve specimens of the bird were ever kept, and all of these came from the Mussoorie or Naina Tal districts of East Punjab and Kumaon between 1500m and 2100m (5000–7000ft) altitude.

With so little observation of the Mountain Quail, there is no way of knowing the reasons for its extinction. It was last seen near Jerepani by a Captain Hutton in June 1868.

PINK-HEADED DUCK

Anatidae
Rhodonessa caryophyllacea
EXTINCT *c.* 1942
Upper Bengal, India

Half a century later, elsewhere in northern India, another distinct monotypic species became extinct. This was the Bengali Pink-Headed Duck (*Rhodonessa caryophyllacea*) which bred largely in the region north of the Ganges and west of the Brahmaputra. Its habitat was tall grassland, floodplains and small lakes and ponds: an area with a small human, but large tiger and crocodile population. In Bengali this waterfowl was called 'Saknal', in Hindustani 'Golabi-sir', 'Umar' in Tirhoot and 'Dumrar' in Nepalese.

Once thought to be a true freshwater duck (of the genus *Anas*), it was found to have sufficient anatomical differences to be assigned a separate genus (*Rhodonessa*). Its bill was reddish-pink, eyes red-orange, and its feet dark with a red tint. It was about the size of a mallard (60cm, 24in.), but was more slender, weighing about 1kg (2lb).

It was not only unique in having a pink head and neck, but was also the only duck to lay perfectly spherical eggs. These eggs were said to be 'as extraordinary as the bird' – so round and white they looked 'like unpolished billiard balls'. In June and July, five to ten eggs were laid in well-formed nests of dry grass.

Hunters reported that these ducks would practice the 'most elaborate affectations of injury to decoy intruders from (their) young'. The call was described as 'loud quacking', while the drake made a 'mellow, two-syllabled "waugh-uh"'. Its flight was easy, rapid and strong, and its wings made a very characteristic clear and soft whistling sound.

A Mr Turner shot a drake close to Shahjahanpur in February 1923 (they were virtually extinct by then): 'The first couple of shots put up about a thousand birds and a flight of white-eyed pochard swung round over me, amongst them being a duck which appeared to me to be a pintail by its flight. I picked him out and dropped him and as soon as I saw him on the water, knew what he was; I waded out to him and he made no attempt to get away.' Turner commented also on 'the beautiful shell-pink of the whole underwing', the very long neck and the 'bill lovely pink with very round nostrils'.

The Pink-Headed Duck was said to be plentiful in the 1880s. In the 1890s about half a dozen could generally be seen in Calcutta's winter market. Although as much as 15 rupees was asked for each dead bird, it was more often sold alive, as its main value was as an ornamental creature. It was not regarded as a good table bird.

By 1915, offers of 100 rupees for Pink-Headed Ducks in the markets of Calcutta got no response. In 1924, J. A. Bucknall wrote of the bird's disappearance. He primarily blamed hunters, claiming that the birds were 'shot all the year by those to whom a closed season is a dead letter'. He commented on how many more fowling pieces (shotguns) there were than 40 years previously and added: 'Then, too, vast areas of swampy ground have been brought under cultivation within

Pink-headed Duck *Rhodonessa caryophyllacea* Extinct *c.* 1942

the last century.'

Strangely enough, the last Pink-Headed Ducks probably did not die in India, where the last recorded specimen was taken in 1935; for captive birds lived on in Clères, France, and Foxwarren Park, England. There were as many as ten of these extravagant-looking ducks in Foxwarren Park for nearly ten years. But although conditions were thought to be ideal, they never bred. The last captive bird died in the English park in 1924.

COUES' GADWALL

Anatidae
Anas strepera couesi
EXTINCT *c.* 1874
Washington Island, Central Pacific

Coues' Gadwall
Anas strepera couesi

In 1874, Dr Thomas H. Streets, an assistant surgeon in the US Navy, discovered a native duck on the tiny Pacific attol called Washington Island. This island, just six degrees north of the equator and south of the Hawaiian Islands, is an attol raised just above sea level. It has two peat bogs and one lake where the raised lagoon has turned fresh in the 250cm (100in.) of rainfall per year. Streets found the 'lake and peat-bogs were tenanted by a diminutive species of duck'. This was later named Coues' Gadwall, or the Washington Island Gadwall (*Anas strepera couesi*). Streets collected two specimens on his visit and none have ever been procured since.

Streets reported that both his specimens were immature: 'the adults, it is presumed, will show the peculiar vermiculated appearance of [the Common Gadwall]. They resemble the immature condition of [the Common Gadwall] . . . but the [Washington Island Gadwall] is immediately distinguished by its greatly inferior size, which hardly exceeds that of a teal, the different colour of the bill (black) and feet, and the singular discrepancy in the lamellae of the bill, which are much smaller, and one-third more numerous.'

Virtually nothing beyond Streets' observations and specimens is known about this bird, and consequently nothing can be said definitely concerning its disappearance. It is known that residents of Fanning Island (about 128km, 80 miles distant) during the late 1800s regularly hunted large numbers of migrant ducks each year on both Fanning and Washington Islands. If these birds were indeed non-migratory residents such a restricted territory could well have allowed them to be extirpated by hunters in a very few years. Whatever the case, there are no reports of their presence beyond the date of their discovery in 1874; and they were certainly gone by 1890.

LABRADOR DUCK

Anatidae
Camptorhynchus labradorius
EXTINCT *c.* 1875
New England and Canadian Maritime Provinces

The only species of waterfowl to become extinct so far in North America is the rather mysterious Labrador Duck (*Camptorhynchus labradorius*). This was a distinctive-looking sea-duck once inhabiting eastern Canada and the United States; however there is no real evidence that it really lived or bred in Newfoundland or Labrador itself. The male birds were black, with white heads and markings; the females were brown above and greyish-white below, with white markings on head, throat and wings. The bills were black with an orange base, and the feet grey-white and black. Because of the striking appearance of the male bird, the Labrador Duck was also called the 'Pied Duck' and even the 'Skunk Duck' in New York, where it was commonly shot with other waterfowl.

Nothing is really known of this bird's behaviour, although the unusual beak, with its exceptionally numerous lamellae and swollen cere, suggests specialized feeding habits. It was a strong flyer and

Labrador Duck *Camptorhynchus labradorius* Extinct *c.* 1875

a good diver, and as with most sea ducks, it was not a good table bird – its flesh tasting strongly of fish. Like the Pink-Headed Duck, the Labrador Duck is considered sufficiently distinct to be classified in a genus to itself (*Camptorhynchus*). The only eggs ever reputed to have been collected were placed in the Staatliche Museen in Dresden, but there is some doubt as to their authenticity.

Despite rumours of the Labrador Duck's northern breeding grounds there is no solid evidence to suggest the bird was found anywhere north of New Brunswick. It certainly did inhabit the sandy bays and islands from the Bay of Fundy to New Jersey, where it was hunted in winter and autumn. In view of its extinction, it is likely that the bird's range was narrower, and its population smaller, than originally believed. Consequently, the fact that it appeared in considerable numbers in the markets of New York and Baltimore during the 1860s suggests that, relative to its population, it was, like the Pink-Headed Duck, simply overhunted.

Another factor may be that the Labrador Duck's population was in fact considerably larger a century earlier, but was dangerously depleted by the same ruthless feather hunters who exterminated the Great Auk on the Atlantic Coast. It is known that feather hunters brought the waterfowl of New England to 'commercial extinction' by 1860, when the birds were so reduced that hunting was no longer profitable and the trade ceased. So, although the Labrador Duck was not scientifically named and classified until 1788, it is not unlikely that the majority of its population had already been destroyed by that time.

All the factors in the bird's demise, cannot now be known or measured. Only one Labrador Duck was ever seen after 1871 – a bird which was shot on Long Island, New York in 1875.

KOREAN CRESTED SHELDUCK

Anatidae
Tadorna (Pseudotadorna) cristata
EXTINCT *c.* 1916
Korea and Japan

Another exotic and colourful casualty in the duck family was the Crested Shelduck of Korea and Japan (*Tadorna* [*Pseudotadorna*] *cristata*). There are only three extant specimens of this particoloured duck (two in Tokyo and one in Copenhagen). The Crested Shelduck was the size of an exceptionally large mallard, and presumably shared all shelducks' (genus *Tadorna*) burrow-nesting habits. However, it was easily distinguished from other shelducks by its crested head – a green crest and grey face in the male, and a black top and white face in the female. Metallic green feathers covered the breast and lower neck in a wide band. In the male the wing coverts were white and the outer secondaries green. The white face of the female bird had black, spectacle-like rings around the eyes. The back and underside of the female were a paler brown than the males, but both were etched with white vermiculated lines.

Although the Crested Shelducks were described and drawn by the Japanese before 1850, the first specimen to be seen by Europeans was collected in 1877. This was a female and is now in Copenhagen. It was not until 1914 that a male specimen was obtained, and in 1916 the third and last bird (another female) was shot. These specimens were taken in Korea and sold to the Kuroda collection, Tokyo. Although a couple of unconfirmed sightings were recently claimed, extensive searches failed to provide any evidence of the Crested Shelduck's survival.

Korean Crested Shelduck
Tadorna (Pseudotadorna) cristata

AUCKLAND ISLAND MERGANSER
Anatidae
Mergus australis
EXTINCT *c.* 1910
Auckland Island, New Zealand

Because of human interference, direct or indirect, many species of New Zealand's waterfowl are now seriously endangered. However, the Auckland Island Merganser (*Mergus australis*) is the only native waterfowl to have suffered extinction since the arrival of Europeans.

In fact, the Auckland Island Merganser's decline seems to have begun with the coming of the Maoris who hunted it on South Island. By 1839 it was only found on the Auckland Islands, where Hombron and Jacquinot, the naturalists attached to the *Astrolabe* and *Zelle*, 'discovered' it.

By this time, the mergansers were very likely in a serious state of depletion. Pigs had been imported in 1806, sealing and whaling stations had been established, and rats and mice from frequent wrecks had long preceded these. As mergansers nest in holes in trees, and sometimes on the ground, these predators, added to outright shooting of the birds, and the later appearance of cats and dogs, had predictable results on the then dangerously localized species.

This distinctive merganser with its thin, serrated beak ('saw-bill') and crested brown-red head, throat and foreneck, could not be mistaken for any other bird on the islands. It fed on small fish and fresh-water shrimps in inlets, streams and estuaries.

Mergansers (genus *Mergus*) are fish-eaters with narrower bills than other ducks, and of the seven species of merganser, the Auckland Island form had the narrowest bill.

Only about 20 specimens of the Auckland Island Merganser exist, the last being collected by Lord Ranfurly for the British Museum in 1902. Several searches throughout the islands have been made for this bird since that time, the latest in 1973, with no results.

Auckland Island Merganser *Mergus australis*

Huias *Heteralocha acutirostris* Extinct *c.* 1907

Exotics and Eccentrics

WATTLEBIRDS TO WOODPECKERS, CUCKOOS TO COURSERS

HUIA

Callaeidae
Heteralocha acutirostris
EXTINCT *c.* 1907
North Island, New Zealand

At the turn of the century, the Duke of York – the future George V of England – made a royal visit to New Zealand where a Maori chieftain ceremoniously presented him with long, elegant black and white feathers. It was the tail-plume of that country's most eccentric bird, the Huia (*Heteralocha acutirostris*). In Maori civilization, the Huia feather had been for centuries the traditional symbol of rank and authority. The Duke promptly placed the feathers in his hat – and by that single action, many believe, he condemned the already threatened Huia to extinction.

The Duke set an immediate fashion for Huia feathers, traditional feathered Maori head-dresses, and stuffed birds. Hunters, curio collectors and ornithologists killed or captured the birds themselves or hired Maoris to hunt them. Traditionally, it is claimed, Maoris were careful not to overhunt this bird; and when its numbers in an area dropped below a certain number, a self-imposed ban was declared by the chieftains. However, in their broken and dispirited state after the war with the Europeans, many Maoris had no alternative but to take payment for the hunting of the Huia.

The famous New Zealand ornithologist, Sir Walter Buller – who himself killed large numbers of Huias – wrote that in one month a small hunting party of Maoris collected 646 Huia skins. The rare bird could not hope to hold out long against such an onslaught.

The Huia was a large (48cm, 19in.) black bird with a broad white band at the tip of its tail. It also had a distinctive bright orange 'wattle' at the base of its ivory beak. Deriving its name from its characteristic cry, it was considered unique among the world's living birds in the extreme structural differences of beak forms in male and female. The difference was so great, that the first European observers believed the male and female made up separate species.

The male bird's bill was short and strong and shaped rather like that of a starling. The female's was long and curved like a nectar-feeder's. The birds were always observed in pairs when feeding: the male bird chopping and breaking up rotten bark in search of insects (primarily the Huhu grub), and the female following and probing crevices the male could not reach. Obviously their bills allowed co-operative, non-competitive feeding between the sexes, yet scientists are at a loss to explain how the Huia, alone of the world's birds, could have evolved such a degree of sexual dimorphism.

In fact the Huia belongs to a most eccentric family endemic to New Zealand. These are the so-called Wattlebirds (Callaeidae), a family of only three species; each species being assigned a separate genus, and each recognizable by the orange or blue 'wattle', a skin excrescence at the base of the beak. Besides the extinct Huia, there are the 46cm (18in.) Kokako or Wattled Crow (*Callaeus cinerea*) and the 25cm (10in.) Tieke or Saddleback (*Creadion carunculatus*) – both are extremely endangered today.

Though trophy-hunters wiped out the last viable breeding population of the Huia, they had not been the major cause of its decline. Introduced predators and forest destruction were the major factors. However, there is one further pressure which, although unproven, quite likely had a considerable effect: disease. Examination of museum specimens has detected parasitic African and Asiatic ticks which were probably transported with the Mynahs that Europeans introduced to the Huia's North Island habitat. It seems likely that these ticks might have carried diseases which would prove as fatal to the Huias as avian malaria was to the Hawaiian Honeycreepers. The last Huia was sighted on 28 December 1907.

AMERICAN IVORY-BILLED WOODPECKER

Picidae
Campephilus principalis principalis
EXTINCT *c.* 1972
Southeast USA

Among the most striking, yet widespread of the world's birds, are the giant woodpeckers. They are large, conspicuously marked birds which need considerable tracts of mature timberland to sustain them. Destruction of such timberland during the last century has been all but universal. Added to this, such beautiful birds have also lost considerable numbers to hunters and collectors.

Among the largest and most famous of all woodpeckers was the American Ivory-billed Woodpecker (*Campephilus principalis principalis*). Although by 1900 it was even rarer than the California Condor, a great deal is known about the Ivory-bill – a 50cm (20in.) long bird, so striking that James Audubon called it the 'Van Dyke'. It became the most sought-after and investigated of all rare birds. There are over 400 traceable museum specimens of the Ivory-bill. James T.

American Ivory-billed Woodpecker *Campephilus principalis principalis* Extinct *c.* 1972

Tanner's classic 1942 study (sponsored by the National Audubon Society) gives us extensive knowledge of its range, habits and needs. The strength and vigour of these birds often astonished even the most knowledgeable ornithologists. Alexander Wilson wrote of one he wounded and kept in his hotel room in Wilmington, North Carolina: 'In less than an hour I returned, and, on opening my door he set up a distressing shout, which appeared to proceed from grief that he had been discovered in his attempts to escape. He had mounted along the side of the window, nearly as high as the ceiling, a little below which he began to break through. The bed was covered with large pieces of plaster, the lath was exposed for at least fifteen inches square, and a hole large enough to admit the fist, opened to the weatherboards; so that, in less than another hour he would certainly have succeeded in making his way through.'

The American Ivory-billed Woodpecker was evidently never a common bird, feeding as it did only in extremely mature forests and requiring a territory of 2000 acres per breeding pair. However, its territory was broad, if specialized and thinly populated. Before the heyday of the timber industry in the American South (1880 to 1900) the Ivory-bill inhabited virtually all mature riverbottom timberland in southeastern North America, and its beak was a valued item of Indian trade. In 1939 a beak was found in an Indian grave as far north-west as Colorado. Mark Catesby, in his 1731 *Natural History of Carolina* . . ., wrote: 'The bills of these birds are much valued by the Canada Indians, who make coronets of them for their Princes and great warriors . . . (and) purchase them of the Southern people at the price of two, and sometimes three buck-skins a bill.'

By 1939, Tanner estimated there were no more than 22 Ivory-bills in the United States, and this figure was based only on the theoretical capacity of breeding areas still intact. Although there is a record of a possible sighting as late as 1972, by the 1960s the population in all of America was estimated at not more than six breeding birds. With no possibility of any real protection or a captive breeding program, the bird must be considered effectively extinct.

There is little data on the other Ivory-bill subspecies (*Campephilus principalis bairdii*) of Cuba. Some ornithologists suggest that there may be as many as 20 extant birds, but most believe the Cuban Ivory-bill is also extinct through forest destruction.

The case is much the same for the Imperial Woodpecker (*Campephilus imperialis*) which, at 55cm (22in.), is the largest of the world's woodpeckers. Its territory is the mature mountain forests of Mexico's Sierra Madres. The possibility of survival for the Imperial is even less than for the Cuban Ivory-bill.

GUADALUPE FLICKER

Picidae
Colaptes cafer rufipileus
EXTINCT *c.* 1906
Guadalupe, Mexican Pacific

Another beautiful American woodpecker was the Guadalupe Flicker (*Colaptes cafer rufipileus*) which lived on the small Pacific island of Guadalupe off the Baja California coast. This island has suffered extreme habitat destruction through the introduction of goats, as well as the introduction of such predators as cats and rats. One collector commented on the bodies of many birds mauled and dismembered by the feral house cats.

The Guadalupe Flicker was a large red-winged, red-tailed and red-cheeked woodpecker with a grey and white body. As with most

Guadalupe Flicker *Colaptes cafer rufipileus* Extinct c. 1906

flickers, its bill was less powerfully built than the larger woodpeckers' (certainly much smaller than the 7.6cm (3in.) Ivory-bill's), but its habits were similar – feeding on smooth-skinned caterpillars and beetles and ants. During the first three-quarters of the nineteenth century, the bird was not uncommon in the restricted area of the large cypress grove on Guadalupe, but was fairly rare among the palm trees on other parts of the island.

In 1886 the naturalist W. E. Bryant found the birds were dwindling with the expansion of the herds of goats. However, in January he observed behaviour that he believed unique to the genus: 'A pair of Flickers were perched facing each other upon a gnarled root about 3 feet from the ground, their heads within a foot of one another. Suddenly the male, who had been sitting motionless before the female, began a somewhat grotesque performance, which consisted in a rapid bobbing of the head. In this he was immediately followed by the female. This . . . they repeated alternately a few times . . . The movement resembled more an upward jerk of the head than a bow . . . I could hear a low chuckling sound while these strange actions were in progress.' Unfortunately, Bryant's behaviour after this observation was typical of nineteenth century ornithologists: 'What the outcome of this lovemaking – for such I regarded it – would have been I did not ascertain . . . The fear of losing the specimens – almost the first I had seen – prompted me to fire. The first shot brought down the female. At the report away flew the male, followed by another male . . . They returned while I was still holding the female and thus gave me an opportunity of securing them both.'

In the context of such a rapidly dwindling population, the behaviour of the last known collector of the Guadalupe Flicker was fairly reprehensible. In June 1906 W. W. Brown reported there were 40 remaining Flickers, and yet saw fit to collect twelve of the birds. There was no recorded sighting of the Flickers after that year and presumably the ravages of goats, goatherds and cats finished them off soon after.

RYUKYU KINGFISHER

Alcedinidae
Halcyon miyakoensis
EXTINCT *c.* 1887
Ryukyu Islands, Japan

One of the most mysterious of the world's birds was the Ryukyu Kingfisher (*Halcyon miyakoensis*). The bird is known only by a single specimen taken in 1887 by the Japanese collector, Y. Tashiro from Miyako Island (in the Ryukyu group) about 224km (140 miles) south-west of Japan and 288km (180 miles) north of Formosa. This specimen is partially damaged, but differs from other related kingfishers in having red feet and longer primary feathers, as well as other marking variations. There are no recorded observations, and we do not know its range.

CHATHAM ISLAND BELLBIRD

Meliphagidae
Anthornis melanocephalus
EXTINCT *c.* 1906
Chatham Islands, New Zealand

The New Zealand Bellbirds (*Anthornis*) gained their name from their most beautiful song which was compared to the ringing of small silver bells. As Sir Joseph Banks wrote in 1770: '. . . I was awakened by the singing of the birds ashore, from whence we are distant not a quarter of a mile. Their numbers were certainly very great. They seemed to strain their throats with emulation, and made, perhaps, the most melodious wild music I have ever heard, almost imitating small bells, but with the most tunable silver sound

Ryukyu Kingfisher *Halcyon miyakoensis* Extinct *c.* 1887

imaginable, to which, may be, the distance was no small addition.'

There were reports, too, of Bellbirds having 'regular concerts': 'Apparently the gathering was for the express purpose of singing together . . . each one contributing to the programme.'

There are two bellbirds: both suffered from European contact, but the larger, gifted with a 'richer and fuller' song, is extinct. This was the Mako-Mako or Chatham Island Bellbird (*Anthornis melanocephalus*), 22.9cm (9in.), a dark-green bird with a dark glossy purple or 'steel blue' head. Cats, rats and forest destruction were the main causes of extinction. However, at the end of the nineteenth century, the rare-bird collectors made their contribution.

The collector W. Hawkins wrote in 1896: 'This bird, too, is nearly extinct. I have no difficulty in selling the skins for £1 apiece; so I have sought diligently for them but it is very difficult to get any of them now'. The last recorded sightings of the birds were in 1906.'

RÉUNION FODY

Ploceidae
Foudia (madagascariensis) bruante
EXTINCT *c.* 1776
Réunion

In the eighteenth century the Marquis du Quene wrote of the Fodies of Réunion: 'The male sparrows have red breasts and, when they make love, redder than ever.' The exact nature of this little bird is uncertain: no specimens exist and the only pictorial evidence is a plate in Daubenton's *Planches Enluminées* of 1779 which has aroused much controversy in ornithological literature.

The real question is whether this Fody was unique or whether it was identical to Madagascar's *Foudia madagascariensis.*

While du Quesne clearly implies that the female Fody was not so brightly coloured or red as the male, de Montbeillard, in his *Histoire Naturelle des Oiseaux* of 1778, simply describes the birds as ruddy coloured in general – with wings and tail only slightly brighter. Nor was he positive that his specimens did come from Réunion.

It would seem de Montbeillard was describing the male out of the breeding season, when its lust-inflamed colouring had subsided.

But a further problem is presented by Daubenton's plate, which shows wings and tail dark brown with yellow-rimmed plumes. This need not be incompatible with du Quesne's account (in emphasizing the red breast he might be taken to imply that the remainder of the bird was not so coloured) but does not fit with de Montbeillard's.

The only explanation consistent with all pictorial accounts is firstly, that Daubenton's picture shows a male Réunion Fody out of the breeding season; secondly, that de Montbeillard described the female; and thirdly, that du Quesne's description informs us of the spectacular plumage-alteration in the cock while breeding.

These deductions seem commonsensical enough but many authorities have preferred to dismiss the species. R. B. Sharpe, for instance, thought Daubenton's plate was simply a picture of the Madagascan bird. However, there are distinct differences: the Fody of Madagascar has a back streaked with black and brown and its ground colour is crimson. The back of Daubenton's bird is uniformly shell-pink.

J. C. Greenway questions Daubenton's reliability, points to the fact that de Montbeillard did not actually collect his specimens himself, and concludes: 'It is quite as likely that the foudias of Réunion were introduced from Madagascar as it is that a species was confined to the island'. But the probability of people stocking Réunion with 'sparrows' at that early date seems quite improbable.

Mako-Mako Chatham Island Bellbird
Anthornis melanocephalus
Extinct *c.* 1906

Réunion Fody
Foudia (madagascariensis) bruante
Extinct *c.* 1776

Delalande's Madagascar Coucal
Coua delalandei
Extinct *c.* 1920

Since other islands near Madagascar have their own subspecies of the Madagascan Fody, it seems likely that Réunion would have too – and at that distance a distinctively altered one. It should satisfy all camps if the Réunion Fody were named as a subspecies of the Madagascan bird – ie. *Foudia madagascariensis bruante.*

DELALANDE'S MADAGASCAR COUCAL

Cuculidae
Coua (Cochlothraustes) delalandei
EXTINCT *c.* 1920
Madagascar

Among the largest of the world's cuckoos (and the very largest of the group called coucals) was the 55cm (22in.) Delalande's Madagascar Coucal (*Coua delalandei*). It was an impressive bird: dark blue above and chestnut below. Its throat was white, and its very long tail was greenish, tinged blue with white-tipped outer feathers – not unlike the Eurasian Magpie's.

The coucal inhabited the northern half of the humid eastern region of Madagascar, perhaps as far south as Tamatave. The Mission Zoologique Franco-Anglo-Americaine reported habitat destruction of the whole lower forest in that region before 1932. Although the last Delalande's Coucal actually captured by Europeans was in 1834 (by the Frenchman Bernier), there were reliable reports of native trapping of the birds in the first two decades of the twentieth century. These hunters worked for themselves in the forests above Fito and Maraontsetra because the feathers were locally prized.

Local hunting, then, probably had some effect on the Coucals (especially after the introduction of modern firearms), but the main cause of their disappearance was habitat destruction and introduced rodents. The bird probably became extinct about 1920; certainly by 1932 there was no evidence of it.

JAMAICAN PAURAQUÉ

Caprimulgidae
Siphonorhis americanus
EXTINCT *c.* 1859
Jamaica

Jamaican Pauraqué
Siphonorhis americanus

The Jamaican Pauraqué (*Siphonorhis americanus*) was one of the two nightjars endemic to the West Indies. The other, the Least Pauraqué (*'microsiphonorhis'*), survives as a highly endangered species in Hispaniola and the tiny island of Gonave.

The name 'pauraqué' is three-syllabled and described the call of the bird, otherwise rendered as 'goo-re-caw', which would be heard especially around dusk as the little nightjars set out hunting. In the breeding season (May and June) the calls could be heard all through the night and even by day, so the bird's presence was never a matter of doubt. Actually seeing it, though, was far more difficult, with its well-concealed nests and its 'deadwood' plumage.

When flushed from its nest or perch, the Pauraqué had a characteristic way of quivering its wings and, apparently, its whole body, from which the Haitian form got its nickname 'Grouille-corps' or 'shaking-body'. James Bond, the leading West Indian ornithologist, describes how they 'would flit a short distance through the scrub like large moths and either settle again on the ground, or . . . fly up into a nearby bush'.

A very small, dark nightjar, the Jamaican Pauraqué can never have been a common bird. Only three specimens were ever collected: two in Westmoreland in western Jamaica and one at Freeman's Hall in Trelawney. This last was shot in 1859.

Although the mongoose has been blamed by some authorities for this bird's disappearance, this is surely an error as the last specimen was taken thirteen years before that destructive viverrine was introduced to Jamaica in 1872. It is more likely that rats were the culprits. In any case the disappearance of the Jamaican Pauraqué was a great loss. It was not a mere subspecies of a mainland family, but probably one of the earliest immigrants to the West Indies. It was thus the most long-established and strongly differentiated of the Caribbean night-birds.

JERDON'S DOUBLEBANDED COURSER
Glareolidae
Rhinoptilus bitorquatus
EXTINCT *c.* 1900
Eastern Ghats, India

Jerdon's Doublebanded Courser (*Rhinoptilus bitorquatus*) was a very beautiful, sleek, plover-like bird (27cm, 10.5in.) with longish legs and an upright stance. It had buff, brown and white plumage. First described in 1848, it was believed to have been a permanent resident in the restricted range of the Eastern Ghats of India.

The white bands for which it was named marked both head and body. There was one that ran spectacularly from the base of the bill, below the eye, down the side of the neck and then curved across the throat, joining the one on the other side in a necklace. The other main band went from the shoulders across the breast in a broad white belt. There was no noticeable sexual dimorphism in the species.

Jerdon's Courser always seems to have been a rare bird. However, it was observed during the 1860s as being found 'in small parties, not very noisy, but occasionally uttering a plaintive cry'. Its habitat was 'thin forest or high scrub, never in open ground'. Compared with other coursers', this bird's flight was very rapid.

So little is known of Jerdon's Courser that the causes of its disappearance cannot now be learned. The last authentic record of the bird was in 1900. Many thorough searches, including one in the 1960s by Salim Ali, the Indian authority, have failed to locate it. It must be presumed extinct.

Jerdon's Doublebanded Courser *Rhinoptilus bitorquatus*

Part Two

MAMMALS

Newfoundland White Wolves *Canis lupus beothulus* Extinct *c.* 1911

The Wolves

WOLVES AND 'WOLF-LIKE' PREDATORS

Today, the wolf is a phantom of dream and fairy-tale: from Red Riding Hood to the werewolves of folklore and bad Hollywood films. The wolf has become a psychological reality that is far removed from the animal's physical manifestation. Werewolf cults can be traced back at least to classical Greek times and have always been linked with witchcraft and lycanthropy – the transformation of men into animals. The wolf of folklore possesses powers of hypnosis, shape-shifting and vampire-like blood sucking. Magical powers are often ascribed to the hair and teeth of the wolf, and its bite was often supposed to be poisonous. In Norse mythology the apocalyptic end of the world comes when a giant wolf devours the sun. 'For what can we mean by the Wolf except the Devil?', asks a twelfth-century bestiary, a medieval Christian book of natural history. In essence, the twentieth-century perspective is not far removed from this view.

The 'true' or Grey Wolf (*Canis lupus*) is, of course, not extinct, but in the last 75 years 7 of its subspecies have been exterminated. Four other 'wolf-like' canids were also hunted into extinction.

Despite human superstitions about the wolf, and the persecutions

which they engendered, the Grey Wolf is an extraordinary survivor. In the face of the most intensive and prolonged attempts at intentional extermination practiced on any creature, the species survives. Among the most adaptable of predators, the Grey Wolf was once found everywhere in the Northern Hemisphere. The wolf can be seen on the ice 500 miles from the north pole and in the blazing heat of the Gobi desert. It can live in tundra, mountain, swamp, forest, desert and grassland. Had the Grey Wolf been less adaptable and widespread, the species would now certainly be extinct.

Very few people would confuse the massive 77kg (170lb) white or slate-blue Arctic Tundra Wolf with a 20kg (45lb) grey-brown Arabian Desert Wolf. Nonetheless these two animals are both classified as *Canis lupus* – although of different subspecies. The Grey Wolf's colouring may vary from black to white to brown to yellow. Similarly, popular ideas of what a wolf is, run a considerable gamut from the Grey Wolf and its numerous subspecies, to a wide variety of unrelated wolf-like creatures: the endangered South American Pampas Maned Wolf (*Chrysoscyon brachyurus*), the rare Ethiopian canid called the Abyssinian Wolf (*Canis simensis*), and the extraordinary Tasmanian Pouched-Wolf (*Thylacinus cynocephalus*) – a carnivorous marsupial which was far more closely related to the kangaroo than to any canid.

In England the last Grey Wolf was hunted down in 1500. In the more remote parts of Ireland and Scotland it survived as late as 1750. Today there are no wolves in France, Belgium, the Netherlands, Switzerland, Germany or Denmark. Small isolated populations of the European Wolf (*Canis lupus lupus*) do still exist in the mountains of Iberia and the Balkans, Greece and Turkey. Amazingly, about one hundred still survive in the Italian Appenines.

The status of the wolf in Asia is difficult to determine because the Russian government is generally unsympathetic to such surveys. Grey Wolf subspecies do survive precariously in northern India (*Canis lupus pallipes*) and the Arabian peninsular (*Canis lupus arabis*), and we know there are viable populations of Asian Tundra Wolves (*Canis lupus albus*), Central Asian Steppe Wolves (*Canis lupus campestris*) and Chinese Wolves (*Canis lupus laniger*). The Yeso Japanese Grey Wolf (*Canis lupus hattai*) was exterminated in Japan but has been reported surviving marginally in Sakhalin in Russia.

Of the more than 30 subspecies of Grey Wolf, just over 20 lived in North America. Because wide-scale hunting of the wolf is recent history there, we have a fuller picture of the Grey Wolf on that continent. We also have a clearer view of its grim and rapid destruction. Wolves have never had dense populations. It is therefore astounding to find that in the 50 years between 1850 and 1900 an estimated two million wolves were shot, trapped and poisoned. Today only one per cent of that number survive, and seven races, about one third of all subspecies in North America, have become extinct.

At the beginning of the nineteenth century, the American Great Plains was one of the richest wildlife areas in the world, yet by 1880 every wolf had gone from that vast land area, along with almost every buffalo, bear, antelope, whooping crane, prairie chicken and in fact any other wild creature big enough to be shot or poisoned. By 1945, aside from the Mexican Wolf (*Canis lupus baileyi*) and a few isolated wolves in Montana's Glacier National Park, the wilderness of north-east Minnesota and Michigan's upper peninsular, there were no Grey Wolves south of the Canadian border.

The wolf-hunter of the seventeenth century might, in an entire lifetime, succeed in killing two dozen wolves, if he was a very skilled hunter. In the nineteenth century, a novice wolfer with a sack of strychnine sulphate might kill as many as 500 wolves in a season. Strychnine was considered the best of weapons against the wily and suspicious wolf, which had previously proved extremely difficult to poison. This was partly due to its intelligence, and partly because of its ability to empty its stomach by vomiting at the first sign of nausea or irritation. Thus, slower-acting poisons like arsenic were fairly inefficient.

It took very little skill to be a wolfer in the nineteenth century. One simply went out and shot a buffalo or some other large animal, filled it with strychnine and came back in a day or so to collect the dead animals. This method of wolf-killing was totally unselective, and the amount of poison spread across the North American continent by these ruthless men was astonishing. Not only did wolves die from the poison, but so did anything that ate meat: coyotes, foxes, weasels, cougars, bears, skunks, badgers, ferrets, ground squirrels, racoons, eagles, bobcats, ravens and – not infrequently – Indians. Water supplies were sometimes accidentally poisoned, so domestic cattle, dogs and even children also died. Furthermore, the violent vomiting and defecation that occur simultaneously with the deadly spasms, resulted in poison being spread on prairie grass where it might remain toxic for a year or more, killing horses, buffalo and antelope.

Finally, by 1900, this use of strychnine was considered too dangerous although government wolfers still used it. Wolf traps became popular, though again this was not a particularly selective method of killing. For every wolf that was killed with poison or trap, six other animals would incidentally be killed.

The reasoning behind the large-scale killing of the Grey Wolf since 1900 is difficult to fathom. By 1900 very few cattle were killed by wolves and far more sheep were killed by domestic dogs. Yet still the bounties rose. The government sent wolfers into land where wolves had never seen men or cattle. As technology advanced and wolves diminished in numbers, the more pathological the hatred became.

Scores of cases of mutilation and torture of trapped wolves have been recorded. In 1905 when there could be no reasonable fear of wolves in Montana, that state's veterinarian carried out a programme of inoculating wolves with scarcoptic mange. This programme was not particularly successful as the disease spread from wolves to cattle and domestic dogs. Yet the practice continued until 1916. In 1928 in Arkansas, in an attempt to deprive of a home the handful of wolves that remained in that state, citizens burned out several thousand acres of national forest.

The first North American wolf to become extinct was the Newfoundland White Wolf (*Canis lupus beothucus*). This was a distinctive and large subspecies of Grey Wolf commonly measuring 180cm (6ft) in length and weighing considerably in excess of 45kg (100lb). The White Wolf's head was big, but was also slender and narrow-skulled, with massive, in-crooked carnassial teeth. It was pure white with only the slightest ivory tinge on head and limbs.

NEWFOUNDLAND WHITE WOLF

Canidae
Canis lupus beothucus
EXTINCT *c.* 1911
Newfoundland

Newfoundland White Wolf
Canis lupus beothucus
Extinct *c.* 1911

Texas Grey Wolf
Canis lupus monstrabilis
Extinct *c.* 1920

Great Plains Lobo Wolf
Canis lupus nubilus
Extinct *c.* 1926

New Mexican Wolf
Canis lupus mogollonensis
Extinct *c.* 1920

Kenai Wolf
Canis lupus alces
Extinct *c.* 1915

Southern Rocky Mountain Wolf
Canis lupus youngi
Extinct *c.* 1940

Cascade Mountains Brown Wolf
Canis lupus fuscus
Extinct *c.* 1950

In 1842, the government of Newfoundland set a bounty on wolves. As late as 1875 packs were still sighted hunting caribou. By 1900 the packs no longer came together, and the wolves were hardly ever seen. In 1911, a lone White Wolf was shot. This was the last of *Canis lupus beothucus* which was named after the tall Beothuk Indians of Newfoundland. The name seemed appropriate enough: Europeans exterminated that race of aboriginal people a full century before they pushed the White Wolf into extinction.

KENAI WOLF
Canidae
Canis lupus alces
EXTINCT *c.* 1915
Kenai Peninsular, Alaska

The Kenai Wolf (*Canis lupus alces*) was among the largest of all wolves. It lived on the Kenai Peninsula in Alaska and is believed to have been hunted out by 1915. It was commonly 45 to 63kg (100–140lb) in weight and often over 180cm (6ft) in length. It is believed its size may have been the result of adaptation to hunting the extremely large moose of Kenai. Since the animal's extinction, however, smaller Alaskan wolves have repopulated the Kenai's territory.

TEXAS GREY WOLF
Canidae
Canis lupus monstrabilis
EXTINCT *c.* 1920
Western Texas to Northeastern Mexico

NEW MEXICAN WOLF
Canidae
Canis lupus mogollonensis
EXTINCT *c.* 1920
Central Arizona and New Mexico

In 1920, two small wolves became extinct: the Texas Grey Wolf (*Canis lupus monstrabilis*) and the New Mexican or Mogollan Mountain Wolf (*Canis lupus mogollonensis*). Both were usually dark coloured, 27 to 36kg (60–80lb) and 135 to 150cm (4½–5ft). The Texas Grey Wolf lived in the desert and prairie lands of western Texas and north-eastern Mexico; while the New Mexican Wolf lived in dry, mountainous central Arizona and New Mexico.

GREAT PLAINS LOBO WOLF
Canidae
Canis lupus nubilus
EXTINCT *c.* 1926
Southern Manitoba and Saskatchewan southward to Texas

Perhaps the best-known of North American wolves was the Great Plains Lobo Wolf (*Canis lupus nubilus*) which inhabited the Great Plain from south Saskatchewan and Manitoba southward to Texas. Called variously the Buffalo Runner, the Loafer Wolf or just the Lobo, this was once one of the most numerous wolves. It fed on the vast buffalo and antelope herds. It was a medium-sized wolf, measuring about 165cm (5½ft) in length and weighing 34 to 45kg (75–100lb). Its colouring varied but tended toward a light or even white pelage. The Lobo became extinct in 1926.

SOUTHERN ROCKY MOUNTAIN WOLF
Canidae
Canis lupus youngi
EXTINCT *c.* 1940
Nevada, Utah and Colorado

The Southern Rocky Mountain Wolf (*Canis lupus youngi*) which inhabited Nevada, Utah and Colorado, seems to have survived until 1940. It was a medium-large mountain wolf, weighing 36 to 49kg (80–110lb) and measuring around 165cm (5½ft) in length. It was predominantly a light-buff colour.

CASCADE MOUNTAINS BROWN WOLF
Canidae
Canis lupus fuscus
EXTINCT *c.* 1950
British Columbia south to Washington State

The most recent wolf extinction is the Cascade Mountains Brown Wolf (*Canis lupus fuscus*). This was a medium-large wolf of evergreen forest and mountain, like the Southern Rocky Mountain animal, also weighing 36 to 49kg (80–110lb) and measuring about 165cm (5½ft). It was often simply called the Brown Wolf because of its basically cinnamon to buff colour. It was found in the Cascade Mountain region of Washington and in southern British Columbia. The final refuge of the Cascade Mountains Brown Wolf was in southwestern British Columbia, where the last animal died about 1950.

SHAMANU OR JAPANESE WOLF
Canidae
Canis lupus hodophilax
EXTINCT *c.* 1905
Japan

In Japan there were once two kinds of wolf, where now there are none. One was a subspecies of Grey Wolf called the Yeso Wolf (*C. l. hattai*) which, although extinct in Japan, is said to survive marginally in Russia. The other was a 'miniature' Japanese wolf (*C. l. hodophilax*), a creature the Japanese called the Shamanu and which the nineteenth-century French natural historian Temminck, in his *Fauna Japonica*, gave the Latin name *Canis hodophilax*. Although

Shamanu or Japanese Wolf *Canis lupus hodophilax* Extinct c. 1905

the Shamanu is now classified as a subspecies of Grey Wolf, an examination of the British Museum specimen (the only one outside Japan) does explain why Temminck and other early observers felt that it differed so much from other Grey Wolves that it should have a specific classification of its own. Its distinct characteristics suggest that it had long been isolated on Japan.

The Shamanu was the world's smallest 'wolf'. It measured 84cm (2ft 9in.) in length without its 30cm (12in.) – rather dog-like – tail. Its dense, short-haired coat was generally ash-grey, tinged in parts with white, russet and brown – however, some yellowish, brown and whitish-grey skins have been recorded. Most remarkably, however was the Shamanu's height measurement of 39cm (14in.) at the shoulder. Its legs, even in relation to its smallness, were extraordinarily short for any true wolf: wolves being markedly 'tall' canids, this characteristic particularly suggests the animal was more closely related to the wild dogs than to the Grey Wolf.

Small as it was, the Shamanu seems to have been greatly feared. By the Ainus, the aboriginal Japanese people, the Shamanu was called the Howling God because it so often howled for hours from hilltops and mountains. One European traveller by the name of Henry Faulds, in the year 1885, described how every Japanese house in the north had, as well as a notice giving street number and family details, a charm to keep the wolves away from their doors. But man, like most animals, is most dangerous when he feels threatened, and the Japanese fear of this miniature wolf seems to have been greatly out of proportion to its actual threat. Before such an obsession the Shamanu could not hope to survive long.

Like wolves everywhere, the Shamanu was hunted and trapped persistently. Besides the uses of the animals' skin, it was also valued for other reasons. One C. P. Hodgson in his *A Residence in Nagasaki and Hakodate 1859–60* describes how 'wolves were brought to the doors of the omnivorous Europeans and offered for sale'.

The main stronghold of the Shamanu was in Honshu, but they were also numerous in Hokkaido and the Kuriles. In these regions the added incentive of a bounty operated. In Hokkaido the local government of Sapporo paid seven yen for a wolf between 1878–82 and 10 yen after 1888. In 1905 a Shamanu was killed near Washikaguchi in Honshu, and its pelt presented to the European traveller, N. P. Anderson. It was the last the world saw of the Howling God that was the Japanese Wolf.

FLORIDA BLACK WOLF

Canidae
Canis rufus floridanus
EXTINCT *c.* 1917
South-eastern USA

As widespread and varied as the Grey Wolf was in North America, it was not the only species of wolf on the continent. In the south-east and south-central United States there once lived the Black Wolf and the Red Wolf.

The Black Wolf or Florida Wolf was at one time classified as *Canis niger niger*, a species unto itself. The Black Wolf was smaller than the Grey Wolf, but larger and stronger than the more closely related Red Wolf (*Canis rufus*). It was pure black and was a true swamp and forest dweller. It grew to a maximum of 150cm (5ft) in length and had a broader muzzle and longer ears than the Grey Wolf.

However, it must be stressed that these three wolves – the Grey, the Black and the Red – are not differentiated primarily by colour

Florida Black Wolf *Canis rufus floridanus* Extinct c. 1917

The Black and Red Wolves are fairly consistently in keeping with their names, although the Red Wolf is, more accurately, grey-red. But the Grey Wolf has considerable colour variations, as the extinct Newfoundland White and Cascade Brown Wolves once demonstrated. The prime considerations for differentiation of species are in body size, shape and cranial structure. In these ways the Black and Red Wolves differed widely enough from the Grey Wolf to be considered a different species. Since this is a simplification of a complex issue, it should be acknowledged that there is considerable debate about these classifications. Both Black and Red Wolves, in some aspects of appearance, shape of skull, ears and habits, had coyote- rather than wolf-like characteristics. However, the Black Wolf and the Red Wolf were not considered sufficiently different from one another to be considered separate species. Consequently the Black Wolf became classified as part of the species *Canis rufus*, although because of its colour and size, it was assigned a distinct subspecies – *Canis rufus floridanus*.

The Black Wolf once ranged over the whole of Florida, Tennessee and south Georgia. It was common in Alabama in the eighteenth century where it roamed in small packs in mountainous areas, but by 1894 it had been driven out of the big swamps near Baldwin and Mobile. In peninsular Florida the last Black Wolf was killed in 1908. In the 1910s the Black Wolf still survived in some rough hilly areas where, driven by starvation, it reputedly made a number of attacks on domestic cattle. Hunting, trapping and poisoning campaigns were rapidly increased with the inevitable results. In Colbert County in 1917 the last reported Black Wolf was shot, but this was a hybrid Black Wolf that had been bred from the smaller and more numerous Red Wolf.

TEXAS RED WOLF

Canidae
Canis rufus rufus
EXTINCT *c.* 1970
Texas

The Texas Red Wolf (*Canis rufus rufus*) was the smallest of North America's wolves. Its average weight was between 18.1kg and 27.2kg (40–60lb). However, it was proportionately a very tall animal. The Eastern Timber Wolf (*Canis lupus lycaon*), for instance, which is one of the smallest Grey Wolves, may have been 9.7kg (20lb) heavier than the Texas Red Wolf, but both animals measured about 72cm (28in.) at the shoulder.

Thus, the Texas Red Wolf's characteristically tall and rangy appearance, its proportionately large ears and its grey-red or grey-tawny colouring made it easily distinguishable from either the Grey Wolf or the Coyote.

It seems strange that, just as individual Black Wolves, in order to procreate, interbred with Red Wolves, so Red Wolves were soon in the same situation with the still numerous coyote. Even without further depradation by man, within a relatively short time this interbreeding would have resulted in the dwindled Red Wolf's extinction. But depredation by man did continue. The Texas Red Wolf was hunted and persecuted in much the same way as the Black Wolf, but survived to the 1970s in a tiny area in south coastal Texas, until the last of its habitat was destroyed by agricultural and even urban development.

Today in the coastal prairies, marsh and swamp thickets of south-east Texas, near Galveston and the adjacent Cameron Parish,

Texas Red Wolf *Canis rufus rufus* Extinct *c.* 1970

Louisiana, we have the living ghosts of the Red Wolf. These are the remnants of the last remaining subspecies of the Red Wolf – the Mississippi Red Wolf (*Canis rufus gregoryi*) which once inhabitated all of the south-central United States.

There are about 300 animals of this subspecies remaining, similar in size and appearance to the extinct Texas Red Wolf. However, unless extraordinary measures are taken, there is virtually no hope that this last reservoir of the Red Wolf genetic pool will survive. Like the Texas Red Wolf, when on its last legs, these wolves have interbred with the coyote, so that a large part of their population survives as a 'hybrid swarm'. They are neither Red Wolf nor coyote, but a hybrid population. Because of persecution and territorial pressures they have very poor stress resistance. One hundred per cent of the population is severely afflicted with heartworm, intestinal parasites and mange. The pups in particular are severely emaciated due to heavy infestations of hookworm. The result is that the hybrid animal which breeds with a healthy coyote is more likely to have its pups survive than the animal which breeds with another hybrid wolf. Increasingly, the surviving pups display more and more coyote characteristics. So even these animals, the last sick and ghostly remains of the once numerous and widespread Red Wolf species of these prairie and swamplands, will soon disappear, either through disease and starvation or through further inbreeding.

WARRAH OR ANTARCTIC WOLF

Canidae
Dusicyon australis
EXTINCT *c.* 1876
Falkland Islands

In 1833 Charles Darwin, on his famous voyage in the *Beagle*, passed through the Falkland Islands off the southern tip of South America and noted the presence of the only predator on the islands, the so-called Antarctic Wolf (*Dusicyon australis*). While on the islands he collected three skin specimens, two of which were later presented to the London Zoological Society.

The creature was 125 to 160cm (4–5ft) in length and had a large wolfish head, but its legs were shorter than a true wolf's and it measured only 60cm (2ft) at the shoulder. Its body coat was a mixture of brown, yellow and black. Its ears were black, its belly white, and its tail was white-tipped like a fox's. Although wolf-like, it was not a wolf at all, nor was it a large fox as others believed. The Antarctic Wolf, or the Warrah as the islanders called it, was something of an enigma. Not only was this unique species the only predator on the Falkland Islands it was (apart from a small mouse) the only land beast on the island. How this could have occurred is a mystery that will probably never be resolved. How could the Warrah have evolved on these isolated islands independent of any related species? Indeed that it was able to survive by feeding on birds' eggs and hunting birds and sea mammals is, in itself, surprising. One theory proposes that its forbears had drifted over on the ice from Patagonia; while a second suggests that the pre-glacial forests on the Falklands once harboured many species, and that only the Warrah survived.

The first description of the Warrah was recorded by Richard Simson who sailed on the *Welfare* in 1689–90 against the French: 'We saw foxes twice as big as in England, we caught a young one alive, which we kept on board for some months.' However, when the British ship engaged itself in a battle with the French, the Warrah decided a ship was no fit place for a beast and leapt overboard.

Warrah or Antarctic Wolf *Dusicyon australis* Extinct *c.* 1876

As Richard Simson's account was not published, Darwin did not know of it, but he had read and noted the adventures of midshipman John Byron of *The Wager* which was wrecked in the Straits of Magellan in 1741. Byron and the crew took refuge on the Falkland Islands. Byron said the Warrahs were as big as middle-sized mastiffs. His men were much alarmed as 'four creatures of great fierceness resembling wolves, ran up to their bellies in the water to attack the people in the boat' so that they put off to sea again. Later they were frightened again, and they seem to have rather over-reacted by setting fire 'to the tussock to get rid of them'. The result was: 'the country was ablaze as far as the eye could reach for several days, and we could see them running in great numbers.'

In 1764, Dom Pernetty in his *History of a Voyage to the Malouin Islands* reported a similar attack: '... officers of M de Bougainville's suite were, so to speak, attacked by a sort of wild dog; this is, perhaps, the only savage animal and quadruped which exists on the Malouin (Falkland) Islands.' However, Pernetty gives a quite likely theory of the Warrah's behaviour: '... perhaps, too, this animal is not actually fierce, and only came to present itself and approach us, because it had never seen men. The birds did not fly from us: they approached us as if they had been tame.' This tameness, whether motivated by hunger or by curiosity, is well documented by later observers, and contributed greatly to the Warrah's extinction.

De Bougainville himself wrote a considerable account of the Warrah in his *Voyage Round the World*: 'The wolf-fox, so called, because it digs itself an earth and because its tail is longer and more fully furnished with hair than that of a wolf, lives in the dunes of the sea-shore. It follows the game and plans its trails intelligently, always by the shortest route from one bay to another; on our first landing we quite believed that they were the paths of human inhabitants. It would appear that this animal starves for part of the year, so meagre and thin is it. It is the size of a dog, and also barks like one, but weakly.'

Darwin in his *Zoology of the Voyage of the Beagle* quotes these other visitors to the Falklands and then goes on to make his own laconic observations and astute prediction: 'Their habits remain nearly the same to the present day, although their numbers have been greatly decreased by the singular facility with which they are destroyed. I was assured by several of the Spanish countrymen, who are employed in hunting the cattle which they run wild on these islands, that they have repeatedly killed them by means of a knife held in one hand and a piece of meat to tempt them to approach in the other ...

'The number of these animals during the past fifty years must have been greatly reduced; already they are entirely banished from that half of East Falkland which lies east of the head of San Salvador Bay and Berkeley Sound; and it cannot, I think, be doubted, that as these islands are now being colonized, before the paper is decayed on which this animal has been figured, it will be ranked among those species which have perished from the earth.'

Soon after Darwin left the Falklands, the colonial government set a bounty on the animals. Bounty and pelt hunters moved in on the already greatly reduced Warrahs. In 1839 Colonel Hamilton-Smith in his *The Dog Tribe* (in *The Naturalist's Library*) was referring to the 'Falkland Island Aquara Dog' when he saw, in: 'the fur stores of Mr G. Astor in New York, a large collection of peltry which came from the Falkland Islands, where, according to reports that gentleman had received, his hunters had nearly extirpated the species'.

Rather miraculously one specimen reached England in 1868 and was kept for several years in the London Zoo. It had been a survivor, along with three small birds, of a huge collection of animals sent on the packet *Fawn*, joining the mail steamer at Montevideo. The shipment included sea lions, foxes, penguins, geese, wolves, starlings and finches. Typically bad and careless shipping resulted in all these creatures perishing. The man responsible for this collection was named Lecomte. He was the same man who used to send the Zoological Society King Penguins whenever he could get them, which, in its turn, was largely the cause of the disappearance of King Penguins from the Falkland Islands. (Although a shepherd near Dunrose House was responsible for the destruction of the last rookery, which he boiled down to waterproof his roof with their oil.)

As the numbers and threat of the Warrah as a predator decreased, tales of its destructive powers seem to have increased. The old vampire superstition surfaced again, and shepherds made unlikely claims of high sheep killings. They insisted absurdly (as men had before with many other 'wolves') that the Warrah killed only to suck blood from the sheep; falling back on flesh only at times of need. The bounty was raised and hunting intensified once again.

The last known Warrah was killed at Shallow Bay, in the Hill Cove Canyon, in 1876. Darwin's prediction came true faster than he foresaw: the Warrah or Antarctic Wolf was exterminated within his own lifetime.

Warrah or Antarctic Wolf *Dusicyon australis*

Atlas Bear *Ursus arctos crowtheri* Extinct *c.* 1870

The Bears

THE ATLAS BEAR AND THE GRIZZLIES

The Atlas Mountains of Morocco was the last stronghold of Africa's only native bear (*Ursus arctos crowtheri*). The Brown Atlas Bear – as it is generally known – seems to have steadily retreated in territory and dwindled in numbers over a period of more than 2000 years until its final extirpation in the nineteenth century, due to the proliferation of European guns in Morocco and Algeria.

The Atlas Bear's original territory was all of North Africa, and fossilized remains have been discovered in caverns throughout that region. The Greek historian Herodotus spoke of the 'Libyan Bear', as did the ancient Roman writers, Virgil, Juvenal and Martial. The Roman magistrate, Domitius Ahenobarbus brought 100 'Numidian bears' to Italy in 61 BC for the cruel sports of the arena.

However extensive the practice of hunting the Atlas Bear may have been in ancient times, this was probably not the main reason for its disappearance from the greater part of its range. The most obvious cause was simply habitat destruction. North Africa today would be totally unrecognizable to the pre-Christian Romans. At that time, most of North Africa was forest, but as more and more of it fell under

ATLAS BEAR
Ursidae
Ursus arctos crowtheri
EXTINCT *c.* 1870
North Africa

Roman rule, their land was ruthlessly exploited to feed the great consumer economy of the Empire. Its timber supplied the navies of Rome with building materials for many centuries, and timber merchants and colonial landowners cleared vast areas for breeding sheep and goats. The characteristically sandy soil of North Africa, without its covering vegetation, could not withstand these changes and soon eroded. Dune formations began to appear. Once this process started, it could not be stopped – and even today the desert that began with the deforestation of North Africa continues to spread.

A small population of bears seems to have survived the bleak shifting of desert sands over the centuries by retreating to the mountainous and still partially forested regions of Morocco and Algeria. Once firearms became available to the native people in sufficient numbers, even this wary and withdrawn Atlas Bear population was at risk.

The eighteenth-century French naturalist, Poiret, was brought the fresh skin of an Atlas Bear by a Bedouin while in Mazoale, and reported that the beast at that time was not uncommon in the Atlas Mountains. As late as 1830, the Emperor of Morocco is known to have had an Atlas Bear captured alive and sent as a gift to the Marseilles zoological gardens.

It was not until the Englishman, Crowther, brought the bear to public attention by his investigations in 1840 that the scientific community really absorbed the fact of its existence. Edward Blythe of the London Zoological Society wrote to his associates in England of his interview with Crowther in Africa in 1841: 'Upon questioning Mr Crowther respecting the bear of Mount Atlas, which has been suspected to be the *syriacus*, he knew it well, and it proves to be a very different animal. An adult female was inferior in size to the American Black Bear, but more robustly formed, the face much shorter and broader, though the muzzle was pointed, and both its toes and claws were remarkably short (for a bear), the latter being also particularly stout. Hair black, or rather of a brownish black, and shaggy, about 10 to 13cm (4–5in.) long – but on the underparts, of an orange rufous colour – the muzzle black. This individual was killed, at the foot of the Petuan mountains, about 25 miles from that of the Atlas. It is considered a rare species in that part, and feeds on roots, acorns and fruits. Does not climb with facility; and is stated to be very different from any other bear.'

As late as 1867 another Frenchman, by the name of Bourguignat, received reports that there were numerous bears near Edough, and others, in the recent past, in Moroccan and Algerian mountain regions. 'The Animal was said to be small, thickset, and brown, with a white spot on the throat, and to be very fond of honey and fruits.'

It is likely that the Atlas Bear lived on a few decades after Crowther's investigations, but there are no confirmed reports of specimens taken since then. Certainly it did not survive the century.

MEXICAN SILVER GRIZZLY

Ursidae
Ursus arctos nelsoni
EXTINCT *c.* 1964
Northern Mexico

The majestic Mexican Grizzly (*Ursus arctos nelsoni*) was the largest native animal in Mexico. Weighing as much as 318kg (700lb) and often measuring 183cm (6ft) from nose to tail, the Mexican Grizzly was, nonetheless, the smallest of the four acknowledged subspecies of American Brown Bear. Because of its distinctive colouring, it was often called the Silver Bear, or 'el oso plateado', 'the silver one'.

Mexican Silver Grizzly *Ursus arctos nelsoni* Extinct *c*. 1964

The American Brown Bear is the largest land carnivore on earth, and the Mexican Grizzly was the first of those great bears to come into contact with Europeans. This was as early as 1540 when the conquistador Coronado marched from Mexico City to the Seven Cities of Cibola in New Mexico, and on to the buffalo plains of Texas and Kansas. At that time the Grizzly Bear's undisputed habitat was the entire western half of North America, from the Arctic to northern Mexico. The full-length of the Rocky Mountains ran like a spine down the great bear's territory.

The type specimen of the Mexican Grizzly was collected in 1899 by Dr E. W. Nelson in Chihuahua. (The specific name of both the Mexican Grizzly and the Mexican Grey Wolf is *nelsoni*, after the same man.) By the 1930s the Mexican Grizzly had been hunted, trapped and poisoned to such an extent that it vanished from Arizona, New Mexico, southern California and southern Texas. In all of Mexico, it could only be found in the state of Chihuahua in the isolated mountain islands of Cerro Compano, Santa Clara and Sierro del Nido Ranges. By 1960, not more than 30 bears survived, and even though a few private citizens attempted to protect this last handful of Mexican bears, others deliberately set out to destroy them.

From 1961 to 1964 ranchers in this region engaged in an intensive campaign of poisoning, trapping and hunting this surviving population. In 1968, Dr Carl Koford made a three month survey of this territory and reported that he could not find any evidence that a single bear survived this onslaught. Despite one notable claim to the contrary, it is now obvious that the ranchers' campaign achieved its goal – by 1964 the Silver Grizzly Bear was extinct.

The specific classification of the Brown Bear – as opposed to the smaller and more docile American Black Bear (*Ursus americanus)* – is probably the most debated and confusing among mammals. By some systems of classification, the Brown Bears of North America and Eurasia (as well as the extinct Atlas Bear of North Africa) make up a single genus of over 100 species. In 1918, the early authority on the Brown Bear, Clinton Hart Merriam, published his *Review of the Grizzly and Brown Bear of North America* which named 86 species in North America alone. In 1928, H. E. Anthony attempted to revise this to 18 species. Since the 1950s, however, the perspective has shifted to viewing the Brown Bears of the Northern hemisphere as a single highly variable species or super-species – *Ursus arctos*. Under this system, experts propose only four distinct sub-species of Brown Bear in North America: the Kodiak Bear (*Ursus arctos middendorffi*), the Alaska Peninsula Bear (*U a. gyas*), the Grizzly Bear (*U. a. horribilis*) and the Mexican Grizzly (*U. a. nelsoni*).

The range of sizes between sub-species is considerable: the Mexican Grizzly had an average weight of around 217kg (500lb) which is somewhat larger than the European Brown Bear (*U. a. arctos*); while the male Kodiak Bear might weigh between 450kg and 720kg (1000–1600lb) and measure 244–305cm (8–10ft) from nose to tail. However, within a single form, such as the Grizzly (*U. a. horribilis*), the range is almost as great. Some Grizzlies weigh as little as the Mexican species, and some as much as the Kodiaks. In 1881 W. F. Sheard shot a Grizzly in the Sierra Nevada Range which weighed 697kg (1536lb

In the same mountain range in 1854, the legendary hunter 'Grizzly Adams' (alias James Capen Adams) trapped his famous tame bear, Sampson, who was weighed in at 685kg (1510lb).

To give some idea of the degree the Brown Bear has suffered from contact with Europeans, however, it is perhaps more useful to look at Merriam's classifications of 'geographic races'. By Merriam's system there would be nearly two dozen extinctions since 1850.

Between 1850 and 1900, the following forms had become extinct: the Big Plains Grizzly, the California Coast Grizzly, the Sacramento Grizzly, the Navaho Grizzly, the Sonora Grizzly and the New Mexico Grizzly – all but one from the southern US. Since 1900, 16 more of these local races have disappeared: the grizzlies of Mexico, Texas, Arizona, Utah and Southern California; seven named for more specific ranges: the Black Hills, Mt Taylor, Lillooet, Tejon, Klamath, Chelan and Twin Lakes bears; two named for naturalists: Baird's and Henshaw's; one named for an Indian tribe: the Apache; and only one named for a physical characteristic: the Flat-headed Grizzly. Obviously, Merriam's distinctions were unscientific, by modern standards, and (as previously mentioned) only four subspecies are recognized in North America today; but his lists do give some picture of the enormous shrinking of the Brown Bear's range in the face of human colonization.

In terms of simple numbers, the picture is even more striking. In 1937, the US Fish and Wildlife Service took a census of the Brown Bear in the United States – excluding Alaska. They gave these population estimates: Arizona 7 bears, Colorado 10, Idaho 44, New Mexico 3, Washington 9, Wyoming 480, Montana 620, all other states 0. Before 1830, California alone had an estimated population of 5000 Grizzlies. By 1937 it had none, and there were less than 1200 Brown Bears in all the states south of the Canadian border.

KAMCHATKAN BEAR
Ursidae
Ursus arctos piscator
EXTINCT *c.* 1920
Kamchatkan Peninsula, USSR

There is one other, quite mysterious, giant bear which seems to have suffered extinction in this century. This is the 'Black Bear' of the Russian Kamchatka Peninsula (*Ursus arctos piscator*), which, like the isolated Kodiak Bear, seems to have developed a gigantic form. Unfortunately, practically nothing is known of this animal though its Latin name suggests that, again like the Kodiak Bears, it took a heavy toll of salmon in their autumn, upriver migrations.

Dr Sten Bergman of the State Museum of Natural History, Stockholm, Sweden, wrote in 1936 in the *Journal of Mammals*, his 'Observations on the Kamchatkan Bear'. He stated that in 1920 in Ust-Kamchatsk he was presented with 'a pelt which far surpasses every other bearskin I have ever seen. It is asserted generally by the hunters that the very largest bears always are quite black. Besides this they are always short-haired. Malaise has told me that on one occasion he saw a skull of a gigantic bear of the black kind. . . .'

To some degree, Russian sources support Bergman's view that a gigantic species did in fact inhabit Kamchatka as late as 1920. Weights of 653kg (1441lb), 500kg (1102lb), and 685kg (1510lb) were recorded by Russian hunters. However, since 1930, any bear in excess of 227kg (500lb) has been thought exceptional in Kamchatka. What the habits and characteristics of this huge bear were, cannot now be learned, nor can the details of its extinction.

Barbary Lions *Panthera leo leo* Extinct *c.* 1922

The Big Cats

LIONS, TIGERS AND JAGUARS

BARBARY LION

Felidae
Panthera leo leo
EXTINCT *c.* 1922
North Africa

In popular portrayals of the decadence of the Roman Empire, the most vivid scenario to come down to our time is the feeding of Christian martyrs to man-eating lions. In this context, it is perhaps difficult to perceive those lions as victims as well, rather than as the agents and executioners of the Roman persecutors. The chief supply of lions for the arena and gladiatorial combats came from North Africa. These were the great shaggy-maned Barbary Lions, the type-race of the species – *Panthera leo leo* – and the most impressive of all lions. The numbers taken captive by the Romans were staggering – both Julius Caesar and Pompey were known to have shown hundreds of these big cats at a time. In very real terms those lions of the arena were greater victims than the men they ate. Since that time Christians have flourished; whereas, the Barbary Lion is now extinct.

The Barbary Lion was among the largest of all lions, and was particularly distinguished by its great mane, which was larger and more extensive than any other subspecies'. It virtually covered half the body. Richard Lydecker, in his *The Game Animals of Africa* (1908), describes the Barbary Lion: 'Very large, dusky ochery, with

the mane very thick, long, and extending to the middle of the back; and a thick and heavy mane on the underparts.' The male would weigh as much as 227kg (500lb) and measure up to 300cm (10ft) from nose to tip of tail. The female was, of course, both paler and smaller and: 'In the female the inside of the foreleg is white.'

Like the Atlas Bear's, the Barbary Lion's territory was once all of forested North Africa, but human destruction of that habitat and the spreading of the desert, combined with continuous hunting, resulted in a 2000 year retreat; culminating in a confrontation with European guns, and extinction.

In recent historic times, the range of the Barbary Lion was from Tripoli through Tunisia, Algeria and Morocco. The last lion in the Pashalik of Tripoli was killed in 1700. According to Louis Lavauden in his *La chasse et la faune cynégétique en Tunisie* (1932), the last Tunisian lion was killed in 1891 at Babouch and the last Algerian lion was killed in the same year in the region of Souk-Ahras.

Others give dates a decade or more later for extinction in Tunisia and Algeria. In H. A. Bryden's *Great and Small Game of Africa* (1899), Harry Johnston wrote: 'Down to the time of the French invasion of Tunis, in 1881, lions were still found in the extreme north-western part of the Regency, close to the Algerian frontier. . . .' and in 1899: 'Lions still linger here and there in S.E. and S.W. Algeria.' According to Johnston: 'What has brought about the extinction of this animal is less the persistent attacks of French and Arab sportsmen than the opening up of the forests and the settling down of the people since the French occupation. The herds are now so carefully tended that the lion has little or no chance of feeding on them, while the Barbary Stag and the gazelles have in that region become very scarce.'

In the same year (1899), A. E. Pease wrote specifically of the Lion in Algeria: 'The Algerian Lion has become so rare that it may be said to be nearing extinction . . . It lingers only in the country that might almost be described as the Mediterranean littoral zone, though an occasional lion is still shot or tracked in the interior, as far inland as the district of Soukarras . . . So long ago as 1862 Général Marguerite wrote that . . . the average number killed did not exceed 3 or 4 a year . . . Before the French came, the Thoks had encouraged the Arabs to destroy them by freeing the two great lion-hunting tribes, the Ouled Meloul and Ouled Cessi, from all taxes and paying liberally for their skins. The French gave only 50 francs for a skin.

'Between 1873 and 1883 the process of extinction is measured in Government returns. The numbers killed for the whole of Algeria were, in the last six years of this period, 1878, 28; 1879, 22; 1880, 16; 1881, 6; 1882, 4; 1883, 3, and 1884, 1.'

The last stronghold of the Barbary Lion, however, was identical to that of the Atlas Bear: the rugged woodlands of the Middle and Great Atlas Mountains. This remained one of the wildest and least developed areas of North Africa well into the twentieth century. Here it could only have been the proliferation of firearms during the civil wars and the rise of banditry in Morocco that resulted in the hunting down of the last of the great beasts. Although breeding experiments that are attempting genetically to reconstruct the subspecies are now being carried out with lions in the Rabat Zoo, the last true Barbary Lion was reported killed in the Atlas Mountains in 1922.

CAPE LION
Felidae
Panthera leo melanochaitus
EXTINCT *c.* 1865
Cape Colony, South Africa

The Barbary Lion was not the first subspecies of lion to be driven into extinction by man in historic times. The first was the European Lion of Greece and Macedonia, which was extirpated by the second century BC. However, even in the more recent past (specifically, the three centuries with which this book is concerned) the Barbary Lion was not the first to vanish. The Cape Lion (*Panthera leo melanochaitus*) of South Africa was the first African subspecies to become extinct. Its disappearance so rapidly followed contact with Europeans, that it is unlikely that habitat destruction was a significant factor. The formidable Dutch and English settlers, hunters and sportsmen simply hunted it to extinction.

The Cape Lion, like its North African cousin, was remarkable for its size and its mane. It is probable that the males reached the same maximum weights of over 227kg (500lb) and lengths of 300cm (10ft). In the luxuriance of its mane, the Cape Lion was only second to the Barbary Lion – however the Cape Lion's mane did not extend to the mid-back and differed also in being very much darker than the North African beast's. In 1846 Charles Hamilton Smith wrote of the type specimen of the Cape Lion: 'The species is of the largest size, with a bull dog head; large pointed ears edged with black; a great mane of the same colour extending beyond the shoulders; a fringe of black hair under the belly; a very stout tail, and the structure in general proportions lower than in other lions.' However, later writers have suggested that Smith's specimen was likely a captive beast which suffered from rickets, thus accounting for its 'lower' structure.

There was, however, no disagreement about the Cape Lion's mane. There was a mounted specimen in the Junior United Service Club in

Cape Lion *Panthera leo melanochaitus*

London which was shot by Captain Copeland-Crawford R.A. in 1836 near Colesberg on the Orange River. R. T. Pocock, the mammals curator at the London Zoo, examined this specimen in 1931 and wrote an article called 'The Lion's Mane', for *Field* magazine: 'The mane is not only remarkable for its luxuriance, length and extension over the shoulder, but also for its blackness. It is indeed wholly black except for the tawny fringe round the face and a certain amount of the same pale hue low down on the shoulder.

'The elbow-tuft and tail-tuft are likewise big and black, but the belly fringe, long and thick behind, becomes gradually shorter and thicker and gradually disappears in front of the chest.

'The interest of this lion lies in its being . . . the only representative, in this country at all events, of the now extinct race of splendid lions which formerly inhabited Cape Colony. . . .'

The exact limits of the Cape Lion's range are not known, as other subspecies inhabited South Africa. Records show that lions were common near Cape Town itself until the end of the seventeenth century, and up until the nineteenth century were found by hunters 'as soon as they got away from the immediate neighbourhood of Cape Town, especially on the Karoo and in the Uitenhage'.

Of the Cape Lion's habits, Ahuin Haagner, in his *South African Mammals* wrote: 'Their food consists of the larger game, mainly antelopes, of all kinds, but also includes zebras, giraffes, and buffaloes. They will kill the donkeys and cattle belonging to prospecting and hunting parties, and will raid Kaffir kraals when driven to it by hunger.' Wishing to cover the full range of the lion's diet, he concludes: 'Man-eating lions are generally old lions with bad teeth.'

The last Cape Lion sighted in Cape Province itself was killed in 1858, but the last of the subspecies was hunted down and shot by General Bisset in Natal, in 1865.

With the extinction of the Cape and Barbary Lions (and the earlier extinction of the European Lion), we can see how the species has been driven from the extreme north and south limits of its range. However, it has been largely driven from its eastern limits as well.

By 1900, the once numerous Asiatic Lion (*P. l. persica*), renowned and feared since Biblical times, was all but obliterated. About 100 Asiatic Lions survived only because the Nawab of Junagardh gave the beast a sanctuary in India's Gir Forest. Today there are still less than 200 Asiatic Lions in this preserve, but their continued existence entirely depends on the precarious politics of the region.

During the 1930s, 7000 people and 60,000 cattle entered this preserve. The result is that the forest is over-grazed to the extent that the bordering Gir Thar Desert advances into the forest at a rate of .8km ($\frac{1}{2}$mile) a year. Although breeding populations of captive animals may now prevent the gene pool of the Asian Lion from entirely vanishing; if the present situation continues, the Asian Lion in its last natural habitat will have vanished within the next 10 to 15 years.

BALI TIGER

Felidae
Panthera tigris balica
EXTINCT *c.* 1937
Bali, Indonesia

Although the lion is considered 'the king of beasts', it is in fact not the largest of the great cats. The longer-bodied and generally heavier tiger is the biggest cat. Of eight subspecies of tiger, one is extinct and the other seven are endangered. Although by 1900 there was already a considerable depletion of the tiger's range and numbers,

Bali Tiger *Panthera tigris balica* Extinct *c.* 1937

as late as 1930 there were no tiger extinctions and the wild population numbered 100,000. Today there is a maximum of 5000. The vogue for tiger trophies and the commercial market for skins are the main causes of this magnificent cat's disappearance from the greater part of its territories throughout Asia, although deforestation and the destruction of the wild game on which it fed are other factors.

The now extinct Bali Tiger (*Panthera tigris balica*) was the smallest of all tiger subspecies. Ernst Schwarg, in 1912, described the Bali Tiger as similar to the small Javan Tiger (*P. t. sondaica*), but smaller still. It shared the Javan Tiger's short, dense hair, but the ground colour was somewhat brighter, and the light markings a clearer white. Schwarg gave head and body measurements for this animal of 153cm (5ft) and a tail measurement of 58cm (23in.).

At the time of Schwarg's account, the subspecies was considered fairly common in Bali. However, a massive decline in numbers occurred between the two world wars when uncontrolled hunting by Dutch colonials and local people (with recently-acquired firearms) was rife and fashionable. By the mid-1930s, a survey of the Dutch Indies for the International Wildlife Protection Commission revealed: 'A few yet live in West Bali but they are having a hard time because they are much sought after by hunters from Java, so they will certainly disappear within a few years. The species also exists in northwest and southwest Bali.'

The tiger disappeared in a very few years indeed, not just from West Bali, but from all parts. The last known Bali Tiger was a female shot at Sumbar Kima, West Bali on 27 September 1937.

Of the other seven subspecies – the Javan Tiger (*P. t. sondaica*), the Siberian Tiger (*P. t. altaica*), the Caspian Tiger (*P. t. virgata*), the Chinese Tiger (*P. t. amoyensis*), the Sumatran Tiger (*P. t. sumatrae*), the Indo-Chinese Tiger (*P. t. corbetti*), and the Bengal Tiger (*P. t. tigris*) – only the Indo-Chinese Tiger and the nominate race, the Bengal Tiger, number more than a very few hundred. Although all Tigers are now at risk, it seems that the subspecies at either end of the size range are the most seriously endangered animals. After the Bali Tiger, the Javan was the smallest subspecies: it is now the rarest of all cats. There are certainly less than six of these animals left, and no real hope can be held out for their survival. At the opposite end of the spectrum is the Siberian or Amur Tiger which is the largest of all cats. The average male Siberian Tiger may measure over 300cm (10ft) in length and weigh 227kg (500lb); while some exceptional animals have been measured at 335cm (11ft) and 318kg (700lb). Due to human predation, there are less than 200 of these huge cats left.

ARIZONA JAGUAR

Felidae
Felis onca arizonensis
EXTINCT *c.* 1905
South-western USA

The largest cat of the New World is the jaguar. In appearance this cat superficially resembles the old world leopard, but it is a more compact, heavier animal. The distinctive colouring is generally sandy-red, marked with large spots encircled by smaller ones, in contrast to the regular spot markings of the leopard's coat. The jaguar's head is larger, its body thicker and its claws stronger than the leopard's, though its legs and tail are shorter. It is 152–244cm (5–8ft) in length and, except during mating, is a solitary animal.

The jaguar is generally a forest-dwelling, water-loving cat which was once found from the southwestern United States to the tip of

South America. Over such a vast range, it is difficult to determine any accurate population figures, but it seems now to exist in any number only in undisturbed jungle areas of the Amazon Basin. The entire species is considered threatened; almost entirely because of the market for its skins. In 1968 and 1969, the United States alone *officially* imported over 24,000 Jaguar skins.

The Mexican Jaguars (*Panthera onca veracrucensis*, and *P. o. hernandesi*) have been near extinction since the turn of the century, but it was the large American subspecies, the Arizona Jaguar (*P. o. arizonensis*) that vanished entirely at the beginning of this century. Its habitat was the mountainous areas of eastern Arizona north of the Grand Canyon, the southern half of western New Mexico, and north-eastern Sonora. In the early nineteenth century, it could also be found in southern California. The United States authority on mammals, C. Hart Merriam 'was told by an old chief of the Kammei tribe that in the Cayamaca Mountain region in San Diego County it was there known as "big spotted lion" in their native tongue'. W. D. Strong in his *Indian Records of California Carnivores* in the *Journal of Mammals* (1926), wrote how the Arizona Jaguar once inhabited the mountains bordering the Mohave Desert, and that 'old people made a practice of following jaguar and mountain lion trails in order to uncover and eat the deer remains the animals buried'.

The Arizona Jaguar was among the largest of its kind, and was 'distinguished by its flatter, more depressed nasal bones from other subspecies'. The last documented sightings and killings of the Arizona Jaguars were in the Sierra de los Caballes in New Mexico in 1904 and 1905.

Arizona Jaguar *Felis onca arizonensis*

Eastern Bison or Buffalo *Bison bison pennsylvanicus* Extinct *c.* 1825

The Bison and the Wild Ox

AMERICAN 'BUFFALOES', WISENT AND AUROCHS

The most vivid popular image of the American 'Wild West' in the very first days of European contact, is that of the spectacular, almost infinite herds of stampeding 'Buffalo'. And, indeed, the continent's heartland once pulsed with life on an almost inconceivable scale. There were antelope by the tens of thousands; huge gatherings of prairie chickens, grouse, cranes; sky-darkening flocks of pigeons and curlews, bears, wolves, vultures, eagles – but above all, the vast thundering herds of giant American Bison (*Bison bison*).

One traveller named Thomas Farham wrote in 1839 that on the Sante Fe Trail it took him three days to ride peacefully through only part of a single bison herd. He estimated the herd to be standing on a grazing area of 3,500sq km (1,350sq miles) – roughly the area of Rhode Island. The herd was considerably in excess of a million animals. As late as 1871, the famous lawman (and one-time buffalo hunter), Wyatt Earp described a similar scene, again involving more than a million animals: 'I could see, twenty or thirty miles in each direction. For all that distance the range seemed literally packed with grazing buffaloes . . . the prairie appeared to be covered by a

solid mass of huge, furry heads and humps, flowing along like a great muddy river . . . Clear to the horizon the herd was endless.'

At that time the Buffalo (which was ten times the weight of a man) outnumbered the human inhabitants of the entire North American continent. Nine years after Earp's account, the Buffalo had vanished. Only a few handfuls hid out in inaccessable wilderness.

The disappearance of the Buffalo was no real shock to white Americans. The American military particularly encouraged and rewarded those who slaughtered the Bison simply for trophies, tongues or hides. In fact, it had been the actual though unannounced policy of the military and the government for some time actively to seek the Buffalo's extinction. One congressman claimed the Buffalo must go because, 'they are as uncivilized as the Indians'.

The connection with the Indian was not an arbitrary one, but the real reason behind the slaughter. As General Sheridan wrote at the time: 'The Buffalo Hunters have done more in the last two years to settle the vexed Indian Question than the entire regular army in the last 30 years. They are destroying the Indians' commissary. Send them powder and lead, if you will, and let them kill, skin and sell until they have exterminated the buffalo.' Sheridan later told congress it should mint a bronze medal for the skin hunters, with a dead Bison on one side and an Indian on the other.

How many Buffalo were slaughtered on the Great Plains cannot now be known, but in the 30 years between 1850 and 1880, more than 75 million hides were sold to American dealers. Practically none of these animals was used, as the Indians used them, for food.

The legendary 'Buffalo Bill' Cody claimed the greatest number of Buffaloes killed by any one man. He killed 4,862 in one year. Had these animals been as efficiently marketed as domestic cattle, this individual's efforts could have fed the entire population of San Francisco of that time for nearly two weeks.

By 1880 the great hunt was over, and there was virtually nothing left. However, a few concerned individuals were able to exert their influence to allow the survivors some protection. In 1905, the American Bison Society was formed and both American and Canadian governments sponsored breeding programs to prevent extinction.

Due to the breeding strength of the Buffalo, these programmes have proved quite successful and the world population is now around 30,000. But there were once four subspecies of American bison. Nearly all the survivors are of the nominate and smallest subspecies – the Great Plains Bison (*Bison bison bison*); while the northerly Wood Bison (*Bison bison athabasca*) which was reduced to less than 100 animals by 1960, survives peripherally as an endangered species. Both other subspecies, the Eastern Bison (*Bison bison pennsylvanicus*) and the Oregon Bison (*Bison bison oreganus*) became extinct.

EASTERN BISON OR BUFFALO

Bovidae
Bison bison pennsylvanicus
EXTINCT *c.* 1825
Eastern USA

Of the Eastern Bison, H. W. Shoemaker wrote in his *Pennsylvania Bison Hunt* (1915), that their range was 'between the east and west slopes of the Alleghenies, migrating between the Great Lakes and the valleys of southern Pennsylvania, Maryland and Virginia, to Georgia'. The Eastern Bison was larger than the plains animal, and 'very dark, many of the old bulls being coal black, with grizzly white hairs around the nose and eyes'. It did not have such a large hump, while the legs were 'long without the contrast between the height of the fore and hind quarters seen in more western animals'.

Hunting apart, the Eastern Bison herds were drastically reduced by habitat destruction, especially deliberate fires which left them without grazing lands. According to a New York Zoological Society report by Martin Garretson, the Bison in Pennsylvania had, by 1790, 'been reduced to one herd numbering between 300–400 animals which had sought refuge in the wilds of the Seven Mountains, where, surrounded on all sides by settlements, they survived for a short time by hiding on the most inaccessible parts of the mountains'.

This last Pennsylvania herd was slaughtered in the 'Sink', a great hollow in the White Mountains of Union County, in the dreadful winter of 1799–1800, as they huddled helplessly in the deep snow. The following year a bull, cow and calf were seen in the same county and the bull (shot the next year) was the last known in the State.

A few Eastern Bison survived longer in West Virginia. One was killed near Charleston, West Virginia, in 1815, but no others were reported until 1825 when 'a buffalo cow and her calf were killed at Valley Head, near the source of Tygart's River'. These were the last Eastern Bison. The 'Buffalo' was extinct east of the Mississippi.

OREGON BISON OR BUFFALO

Bovidae
Bison bison oreganus
EXTINCT *c.* 1850
Oregon, Idaho and California, USA

The most westerly subspecies of Bison, the Oregon Bison (*Bison bison oreganus*), fared no better than its eastern cousin. Although it more closely resembled the typical Plains Buffalo, it was a larger animal with wider, straighter horns. It ranged between south-western Oregon and northeastern California to southern Idaho and extreme northern Nevada.

Oregon Bison or Buffalo *Bison bison oreganus*

There is not a great deal of historical documentation on this subspecies, but records do exist of sightings by Europeans in the 1830s in Idaho, while earlier Indian accounts from Oregon and California are supported by finds of bones from entire herds. This evidence indicates seasonal, pasture-seeking migration.

Vernon Bailey, writing in 1936, gave Indian oral accounts of the bison existing in considerable numbers in the Malhew Lake region of Oregon until around 1850. (Again these accounts are matched by large finds of bones in Malhew Lake itself.)

The introduction of horses did not affect the range of the Oregon Bison until the beginning of the nineteenth century, but it is believed that improved hunting and transportation afforded by the horse, and the ready, though distant, market for hides, encouraged the Indian extermination of most of the bison before direct European contact. Vernon Bailey concluded that there was 'no question that only a few generations back buffalo covered in considerable numbers many of the large valleys of southeastern Oregon and that they disappeared after the introduction of horses among the Indians and before many firearms were obtained'.

CAUCASIAN WISENT

Bovidae
Bison bonasus caucasicus
EXTINCT *c.* 1925
Caucasian Russia

The genus *Bison* is not, of course, restricted to North America. It consists of two widely separated species: the American Bison or Buffalo, and the European Bison or Wisent. (Some authorities now regard all Bison as conspecific.) The European Bison once consisted of two subspecies: the surviving Lithuanian (lowland) Wisent (*Bison bonasus bonasus*), and the now extinct Caucasian (mountain) Wisent (*Bison bonasus caucasicus*).

The Wisent is generally a larger, taller, longer-legged animal than the American Buffalo. It often weighs in excess of one ton. It measures 340cm (11½ft) in length (without tail) and 200cm (6½ft) in height. It does not have the American Buffalo's massively heavy forequarters and small hindquarters – such a body structure only being an advantage on flatlands, and a considerable disadvantage in uphill running. (The distinction was even more pronounced in the extinct Caucasian 'Mountain' Wisent subspecies.) Its longer body is less barrel-like and its hindquarters stronger.

The Wisent is lighter (almost fawn-coloured), shorter-haired and less 'wooly' or shaggy-maned than the American Buffalo, but the hair on its hindquarters and tail is heavier. Its head is smaller and is carried higher, and its horns are shorter, thicker and blunter.

The Wisent is a much more wary beast than its American relatives – indeed, it was said to be almost unapproachable in the wild, where it formed large herds only during the mating season. It is a non-migratory, woodland browser, feeding on grasses, ferns, tree bark and acorns and generally living in family groups of a dozen or less.

Both the Lithuanian and the more southerly Caucasian subspecies were once widespread throughout Europe and western temperate Asia, but they retreated with the destruction of the forests and increasing human populations. Wisent were driven out of most of their range by hunting over more than three thousand years. Once the use of firearms became widespread, the last relic populations were speedily hunted down and destroyed. In fact, the Wisent's survival into this century was entirely due to the whim of the Czars of Russia.

Caucasian Wisent *Bison bonasus caucasicus* Extinct *c.* 1925

Only Imperial protection and the threat of severely enforced penalties since 1803 allowed a population of 700 Lithuanian Wisent to survive in Poland, and another 500 or less Caucasian Wisent in the Kuban district of the Caucasus, until 1914. However, World War I and the Russian Revolution proved as fatal to the Wisent as they were to its Imperial protectors. In the turmoil that followed, the gamekeepers disappeared and the Wisent was hunted down with a vengeance. By 1923 *both* subspecies were extinct in the wild.

It was the absolute end for the Caucasian Wisent. Only one bull is known to have been in captivity – and animal called Kaukasus, owned by the Hamburg animal-dealer, Karl Hagenbeck. It died on 26 February 1925, and with its death the Caucasian Wisent was truly extinct.

The Lithuanian subspecies, however, survived by a hair's breadth because the Czar had made gifts of a very few captive bulls and cows before the species was exterminated in the wild. The story of the Lithuanian Wisent after the war is one of the great conservationist success stories. By 1919, the Czar's 700 Lithuanian Wisent in the Polish Bialowieza Forest had been wiped out and the majority of the world's population of Wisent were on the Duke of Hochberg's estate in the Pszcyna Forest in Pless. This herd was descended from the gift of one bull and three cows from the Czar's reserve in 1865. However, in 1921 political upheavals in Pless, as well, resulted in the loss of all but three of these Lithuanian Wisent.

Fortunately, in 1923 the Polish zoologist Jan Sztolcman formed a society for the conservation of the European Bison in Germany. Without this group it is likely that the captive beasts would have lived out their days, breeding very little or not at all, with the consequent extinction of the entire species. However, through the efforts of the society, the three Hochberg Wisent, together with three others gathered from zoos in Germany and Sweden, were brought together. These six animals made up the entire purebred population of the Lithuanian Wisent. Amazingly, the venture succeeded. Although the Caucasian Wisent is extinct, two thousand European Bison in the form of the Lithuanian Wisent (and some mixed-blood Wisent-American Bison) survive today. Many of these are now being re-introduced into the wild with considerable success.

AUROCHS

Bovidae
Bos primigenius
EXTINCT *c.* 1627
Europe

Another giant grazing animal of Europe was the legendary wild ox, the Aurochs (*Bos primigenius*). In prehistoric times the Aurochs inhabited all of temperate Europe and Asia and was hunted by Man from the most ancient times. Within the last two thousand years, however, its range became restricted to central Europe.

The Aurochs were the archetypal wild black bulls of the ancient mythologies: in such a guise the Thunder-god, Zeus, abducted the maiden Europa, from whose name comes the continent's. As the Aurochs' Latin name implies, it was the wild breeding stock from which domestic cattle are descended. Yet, it was said to have been as different from domestic cattle as the wolf is from the dog.

The aurochs were massive beasts, measuring 180cm (6ft) at the shoulder with impressively long and spreading, forward-curving horns. Julius Caesar described them in his account of the Black Forest in *De Bello Gallico*: 'They are but little less than Elephants in size, and are of the species, colour and form of a bull. Their strength is very great, and also their speed. They spare neither man nor beast that they see. They cannot be brought to endure the sight of men, nor be tamed, even when taken young. The people who take them in pitfalls, assiduously destroy them; and young men harden themselves in this labour, and exercise themselves in this kind of chase; and those who have killed a great number – the horns being publicly exhibited in evidence of the fact – obtain great honour. The horns, in amplitude, shape and species, differ much from the horns of our oxen. They are much sought after; and after having been edged with silver at their mouths they are used for drinking vessels at great feasts.'

Many domestic breeds of cattle are claimed to have descended directly from the Aurochs – most notably, the black Spanish fighting-bull, which shares with the Aurochs the characteristic light-coloured line along the spine.

The indiscriminate hunting of the Aurochs continued in Europe until around the tenth or eleventh century, by which time it had been driven from everywhere but pockets of wilderness in East Prussia, Lithuania and Poland. In 1299 the Duke of Boleslaus forbade its hunting in Masovia, as did his descendant Duke Ziemovit in 1359, and after 1410 this region of Poland seems to have been the Aurochs' last refuge. It survived longest in the Jaktorow Forest in West Poland. In 1550 there was still a herd in the forest and in a royal preserve nearer Warsaw. However, in 1599 there were only two dozen animals left. By 1620 only one Aurochs remained alive. It died in 1627, rendering the legendary species extinct.

Aurochs *Bos primigenius*

Blue Bucks *Hippotragus leucophaeus* Extinct *c.* 1799

The Fleet Grazers

ANTELOPES, DEER, SHEEP AND GOATS

BLUE BUCK
Bovidae
Hippotragus leucophaeus
EXTINCT *c.* 1799
Zwellendam, Cape Colony, South Africa

If the Great Plains of North America were astonishing to European eyes for their massed herds of Bison, how much more varied were the savannahs and veldte of Africa with their hordes of zebra, buffalo, giraffe, elephant and, above all, antelopes. Yet nature's prodigality on the Dark Continent was deceptive – even apparently numerous animals might, through undisturbed millenia, have adapted to local and specialized niches which Europeans could destroy almost overnight. The first African animal to disappear in historic times was the Blue Buck or Blaauwbok (*Hippotragus leucophaeus*), a relative of the Roan and Sable Antelopes, which lived only in Zwellendam province of the old Cape Colony and became extinct within eighty years of its discovery. The last specimen was shot in 1799 or 1800.

The first settlers in the Cape called this antelope the 'blue goat' after its curved horns, blue-grey coat and seasonal beard. 'This is the species,' wrote the English naturalist Pennant in his 1781 *History of Quadrupeds*, 'which, from the form of the horns and length of the hair, seems to connect the goat and antelope tribes.' It seems that the settlers valued its skin more than its flesh which, according to the

German Kolbe who first described the Blue Buck in 1731, was 'generally given to the Dogs'. Many skins reached Europe, though virtually none have survived, and it was from some of these that the Russian Pallas derived the first scientific description of the antelope in 1766.

By this time Zwellendam province (in the northwest of the Cape) was becoming populous, so that eye-witness accounts of the Blue Buck in the wild may be descriptions of an animal already reduced and in retreat. It seems, though, that it normally grazed on the open veldt, either in small troops or quite alone, but towards the end of the eighteenth century retreated to the narrow valleys between Stellenbosch and Graaf Reinet. It was not, however, particularly shy or highly-strung, was easily stalked and shot, and had none of the savagery of temperament for which its much larger cousin, the Roan, is renowned.

Its relationship to the Roan and the other *Hippotragus*, the handsome Sable Antelope, has been the cause of some controversy. Many nineteenth-century naturalists (including some who had examined museum specimens) believed that the Blue Buck was simply a dwarf or immature Roan Antelope, lacking the rufous facial and chest markings. In recent years, though, two zoologists (the Swede Sundevall and the Austrian Köhl) have established clear distinctions between the Blue Buck and its two relatives: its horns, its colouration, its form and above all its size separate it. But it is not surprising that such an early extinction, leaving only five museum specimens, should result in some confusion.

The animal's colour is another example of this. Almost all the early accounts of the Blue Buck dwell on its exquisite coat and the effect of death upon it. Pennant's description is typical: 'Colour, when alive, a fine blue of velvet appearance; when dead, changes to bluish-grey, with a mixture of white . . .' And yet our best description of the Blue Buck in life and in death comes from the pen of the French explorer and naturalist Levaillant who shot one in 1781 in Soete-Melk valley and who says emphatically: 'Alive or dead, it looked just the same to me, and the colour of the specimen I brought back has never altered.' But then Levaillant had originally taken the antelope for a 'white horse' before his Hottentot servant identified it as a Blue Buck. We must assume that this 'whiteness' was the result of sun-glare on the extremely glossy pelage, but it leaves even Levaillant's eye-witness account somewhat tarnished.

Perhaps the most convincing hypothesis regarding the Blue Buck's true colour is that put forward by Graham Renshaw, the authority on African quadrupeds, who examined several museum specimens in 1922 and wrote: 'Many antelopes, when age has thinned their coat exhibit a bluish appearance due to the underlying hide: and if we suppose – not unreasonably – that the first "Blue Antelope" shot was one which, by age and infirmity, fell an easy victim to the antediluvian weapons of the colonists, such a specimen would exhibit the "blue velvet" appearance in the highest degree.'

The last reports of the living Blue Buck were made by a local resident, Sir John Barrow, in 1796. He had supposed the antelope extinct but that winter half a dozen reappeared in the wooded hills above Soete Melk. By 1799, (Litchenstein's *Travels* say 1800) the last of these had been shot and their skins sent to Leyden. The Dark

Continent had suffered its first animal casualty.

A far more numerous and widespread family of antelopes is the Hartebeests whose representatives are found the length and breadth of Africa. The most northerly and southerly subspecies, however, are extinct. The Bubal Hartebeest (*Alcelaphus buselaphus buselaphus*) ranged in ancient times from Morocco to Egypt and possibly into Palestine and Arabia too, and the discovery of its horns in Egyptian tombs at Abadiyeh indicate its importance mythologically as well as as a food source. These characteristic, lyre-shaped horns seem to have suggested a relationship with the oxen rather than the antelopes to ancient and modern man equally: Henry Barth, in his 1857 descriptions of the Anahef hills in the central Sahara, wrote of 'large herds of wild oxen (*Antilope bubalis*)' which evaded his party by 'climbing the rocks with much more ease than men'.

He was certainly talking about the Bubal Hartebeest which, standing nearly 122cm (4ft) at the shoulder, were impressive creatures, but the most significant aspect of his account is that it chronicles the Bubal's retreat from the encroachment of Europeans and firearms into the most desolate desert strongholds.

By the start of this century it was only to be found in the southern mountains of Algeria and the Moroccan High Atlas. The large herds found north of the Atlas Mountains a hundred years earlier had vanished, leaving only fond memories in the minds of a few French colonels 'who had shot them in the great battles of game, which massacres were organised in the early days of the French occupation'. But there was no lack of hunters to brave the harsh terrain and

BUBAL HARTEBEEST

Bovidae
Alcelaphus buselaphus buselaphus
EXTINCT *c.* 1923
Algerian desert and Moroccan High Atlas

Bubal Hartebeest *Alcelaphus buselaphus buselaphus* Extinct *c.* 1923

search out the last survivors. The last reliable report of a Bubal in Algeria was of one shot in 1902, though claims have been made of very much more recent sightings. In Morocco the Hartebeest was reported from Missour in 1925, but few authorities credited the claim, and the female which died in a Paris Zoo in 1923 is usually held to have been the last of its kind.

One of the ironies of the Bubal Hartebeest's disappearance was that, despite its desert habitat, it was very amenable to captivity and the company of man. One of the Paris specimens lived in captivity for over 18 years. To a modern conservationist this species would have seemed an ideal subject for a rescue-breeding programme but the chance was missed.

CAPE RED HARTEBEEST

Bovidae
Alcelaphus caama caama
EXTINCT *c.* 1940
South Africa

Cape Red Hartebeest
Alcelaphus caama caama

At the continent's other end another Hartebeest, the Cape Red form (*Alcelaphus caama caama*), was saved from extinction for over a century purely by the concern of one farming family, and when the farm was sold in 1938 the antelope disappeared.

The Cape Hartebeest was abundant throughout South and South West Africa when the white settlers arrived, and formed two subspecies, one of which (*A. c. selbornei*) survives in Namibia. The Red form was so decimated by the Europeans that the 'large troops even in the immediate vicinity of Cape Town' reported in 1800, had vanished within 25 years and only a tiny remnant survived to the north in Natal. A letter in the 1833 report by the International Office, Brussels, says: 'Cape Hartebeest was formerly abundant in the midlands of Natal, but now the farmers have destroyed them all, with the exception of about 25 on a farm owned by Messrs Moe Bros, who do everything possible to protect them against the bloodlust of neighbouring farmers and the savage attacks of dogs owned by the natives.'

In 1938 the herd was still there, now numbering about 55, but the farm was up for sale. Despite some agitation, especially by a Captain G. C. Shortridge, the herd was allowed to disperse, though some animals, more or less hybridized, may be found in zoos and reserves.

The persecution of the Cape Hartebeest by most farmers has an explanation and the antelopes' previously mentioned resemblance to the cattle family is relevant, though the Cape Hartebeest was a smaller and narrower horned creature than the Bubal. The farmers believed that the hartebeests carried the cattle disease, *shotsiekve*, and did everything they could to keep the antelopes off their ranges. They were mistaken in their suspicions but their precautionary measures were so drastic that by the 1940s the Cape Red Hartebeest was extinct and its close relative (*A. c. selbornei*) had been completely eliminated inside South Africa.

RUFOUS GAZELLE

Bovidae
Gazella rufina
EXTINCT *c.* 1940
Algeria

While the Bubal Hartebeest was being hunted down in its thousands by the *colons* of Algeria, a much rarer and more elusive creature was being nudged towards extinction in the same area. The exquisite Rufous Gazelle (*Gazella rufina*) is known to us only through three specimens and retains much of the mystery in death which it held, even for the local natives, in life.

It was a forest dweller, dark coloured, large (152cm, 5ft long) and with thick horns a foot in length. Even compared to the near-legendary gazelle of the Atlases, the Rufous Gazelle was regarded as an enigma.

No one had ever seen the female of the species and even the few hunters who had seen the males disagreed about their habitat. The only people who spoke of it with any confidence were the furriers of Oran who saw it simply as a rare and costly pelt which came into their hands only once every three or four years. By their accounts there were still some gazelles alive in 1925.

In 1929 the French naturalist L. Joleand defined its territory, past or present, as the southern part of the Oran-Morocco border country whence it had vanished at about the same time as the Bubal Hartebeest. Seven years later Joleand's colleague, Heim de Balsac, reported accounts of gazelles in the mountainous forests above the Chelif valley between Oran and Algiers. Balsac believed that small bands of Rufous Gazelle might still be found there, but there have been no subsequent reports. The increasing spread of roads, settlements and firearms in Morocco makes it unlikely that we will ever know more of this strange forest gazelle than we do today.

SCHOMBURGK'S DEER

Cervidae
Rucervus schomburgki
EXTINCT *c.* 1932
Eastern Siam (Thailand)

In other continents the deer family occupies the ecological niches reserved in Africa for the antelopes, and in Asia, Siam (Thailand) harboured in Schomburgk's Deer (*Rucervus schomburgki*) a creature almost as mysterious as the Rufous Gazelle of Algeria. Like the gazelle, this deer was never seen in the wild by a European, though one live specimen was exhibited in the *Jardin des Plantes*, France, in 1867. From its discovery in 1862 to its extinction 70 years later, only 200 skins were exported, and though this toll was hardly enough in itself to account for the deer's disappearance, European interest did add to existing pressures on this rare swamp dweller.

Rufous Gazelle *Gazella rufina*

Its antlers were both its glory and its curse: complexly-tined and curving forward over the stag's head, they made a spectacular trophy. But it was for their supposed medicinal and magical properties that the horns of Schomburgk's Deer were most eagerly sought. As Ulrich Gühler, our best authority, drily puts it: 'The antlers figure particularly in the Chinese pharmaceutical trade.' So avidly were they hunted that one authority (Ziswiller) goes so far as to attribute the deer's extinction to 'religious persecution', which is somewhat misleading.

Habitat change was at least as decisive to the fate of Schomburgk's Deer. In the mid-nineteenth century, there were herds of them in the swamps near Rangsit and during the annual floods the deer were pursued by boat, marooned on small islands and speared. They could sustain their numbers under these traditional pressures, but once swamp-drainage and irrigation changed the habitat, and railroads opened up the land around the swamps, the balance turned against the animals. First they took refuge in bamboo jungle, to which they were not well adapted, and then these too were cleared in low lying areas to make way for rice fields. Though the deer might find the paddies a congenial habitat, they had no cover and their presence was not appreciated by the farmers who by now had firearms.

By 1920 Schomburgk's Deer was virtually extinct, a few stragglers being reported on the Pu Kio range, where the aboriginals were hunting them with dogs; and along the Owe-Noi river in quite dense forest. It was in this latter region, near Sayok, that the last known Schomburgk's Deer was shot by a policeman in September 1932. Though hopes are still expressed that it may reappear there (and even in Yunnan and Indo China where it once occurred) the likelihood is remote indeed. One stuffed specimen and a few skulls and antlers are all that remain of it today.

DAWSON'S CARIBOU

Cervidae
Rangifer dawsoni
EXTINCT *c.* 1908
Queen Charlotte Islands, British Columbia, Canada

Much the same can be said of another wetland casualty, the Dawson's Caribou or Reindeer (*Rangifer dawsoni*) of the Queen Charlotte Islands, Canada's most westerly outpost, just below the Alaska panhandle. This caribou was a very small, pale, possibly relic subspecies, which had a unique habitat for its family. All other species range the dry interior of the north and the whole land area of the arctic region; by contrast the swamp barrens (muskeg) of the Queen Charlottes, where the only firm land is wooded and the humidity often phenomenal, seem quite unsuitable.

The Haida Indians, who had inhabited the islands for at least five thousand years before the Europeans came, did not know about the caribou even by tradition. They did not use the interior of the islands until the white market for furs opened up and even then they stuck to the river banks where they set bear, otter and marten traps. Ironically, all the known specimens of caribou were killed by Haidas – stirred into tracking down the animals by white collectors.

The first published mention of the caribou is in G. M. Dawson's first report on the islands in 1878, when he wrote that there was good evidence for the existence of wapiti in the north part of Graham Island. In 1890 he corrected 'Wapiti' to 'Caribou'.

In 1880 Alexander Mackenzie, a trader for the Hudson's Bay Company in Old Massett, offered a reward to any native who brought

Schomburgk's Deer *Rucervus schomburgki* Extinct *c.* 1932

him in a specimen. A Haida named Elthkeega had some bear dogs and claimed that they had driven a cow and a bull caribou into Virago Sound where he had killed them. He brought the bull to Mackenzie for the reward.

Mackenzie sent out a fragment of the skull with part of an antler attached and it was eventually lodged at the Provincial Museum in Victoria. In 1900 Ernest Thompson Seton, the pioneering naturalist and popularizer, examined it and described it as coming from a new species. He gave it its present Latin name.

But most naturalists were sceptical and white men on the Charlottes, notably the missionaries Collinson and Keen, dismissed the evidence. They believed the natives had brought the animal (or perhaps just the skull) from Alaska to claim the reward.

The 'mackenzie caribou' began to be talked of in scientific circles with the same amused contempt as the 'hastings rarities'. But another missionary, the Revd Charles Harrison, a fluent Haida speaker who had tasted the dried meat of the animal, was far less sceptical about the natives' stories. In fact, now that interest had been aroused, several of them came to tell him of sightings in Virago Sound near the ruined village of Kung.

In March 1901, Harrison himself spent ten days in the woods along the Sound accompanied by the son of the sceptical Revd Collinson and five natives. They saw abundant tracks and dung and they retrieved some Caribou hair. The evidence was mounting. A year later a British naval officer, Captain Hunt, not only saw tracks but brought back a Caribou horn. Mr Harrison concluded that the animal's range was very limited: comprising areas of Graham Island north of latitude 54°, and bordering Virago Sound.

By now the US and Canadian scientists were having to revise their views, and both hunters and naturalists came searching. The most graphic description comes from the pen of the hunter Charles Sheldon in his *Wilderness of the North Pacific Coast Islands*. Sheldon was a humane and learned man for a trophy hunter of that period, but there is a comic ring nevertheless to many of his experiences on the sodden barrens of the Charlottes. His patronizing incomprehension of his Haida 'gillies' is only matched by their own evident contempt for a man whose sole ambition seemed to be to get as wet as possible. His experiences were terrible: he went out on the barrens in October and November before any frost could make the going firmer, and almost every day his diary noted 'another rainstorm'. But he did persevere and he did see signs of caribou, though nothing very recent. Perhaps that was as well for the beasts; after nearly a month in the bush he wrote: 'My caribou hunt had ended. During the whole trip my rifle had not been cocked.'

Sheldon retained his interest in the elusive reindeers and corresponded with the now retired Revd Harrison. And in this way he learnt that the caribou had at last been found – and shot.

On 1 November 1908, two half-breeds, Matthew Yeomans and Henry White, were hunting in a large swamp barren about 5km ($3\frac{1}{2}$ miles) inland from Virago Sound and towards the southern end. They saw four animals near the centre of the barren: two bull caribou with horns, a cow and a calf. The animals showed no fear at all and 'stood quite still until one after another, except the calf, were shot down'. It

Dawson's Caribou *Rangifer dawsoni* Extinct *c.* 1908

seems rather incredible that the population could naturally have shrunk to just four, and that that one orphan calf was the last of its kind; but there have been no reliable sightings since that day and though occasional rumours of animals seen near Naden continue to circulate, it must be borne in mind that both Black-tailed Deer and Wapiti have been introduced to the islands and could give rise to false reports.

GREENLAND TUNDRA REINDEER
Cervidae
Rangifer tarandus groenlandicus
EXTINCT *c.* 1950
Eastern Greenland

Rangifer dawsoni was a form of the Woodland Caribou. The other, larger, branch of the reindeer is the Barren-grounds or Tundra form of which one subspecies (*Rangifer tarandus groenlandicus*) has become extinct this century in eastern Greenland. Some authorities dispute its status, however, and regard it as a form of Peary's Caribou (*Rangifer torandus pearyi*) which is extant in northwest Greenland. Since the range of the two was so widely separate, however, and since the habitats were significantly different, it seems wisest to record this animal's passing about 1950 as an effective extinction and to leave it to zoologists to argue the finer issues of relationship.

EASTERN ELK (WAPITI)
Cervidae
Cervus canadensis canadensis
EXTINCT *c.* 1877
Eastern United States and Canada

The caribou are majestic and graceful animals, but human observers in North America have usually been more impressed by another large native deer, the Wapiti or – as it is more often, if inaccurately, called – 'Elk'. The true elk is the Moose, and the wapiti is really only a giant form of the European Red Deer which seems to have gained in grace from its increased stature (183cm, 6ft at shoulder). Two of the world's eleven subspecies of wapiti are extinct and since they are both American forms their vernacular name of elk is retained.

The Eastern Elk (*Cervus canadensis canadensis*) were creatures of the open plains, very numerous and as important to the plains Indian cultures (especially the Sioux or Lakotah peoples) as the buffalo. They provided meat, leather for robes and tepees, thongs and ropes and – a cosmetic feature which proved their undoing – highly prized canine teeth from their upper jaws. As one authority (G. M. Allen) has pointed out, the enormous range of the Eastern Elk could be mapped 'by plotting the cities, counties, creeks and rivers named after it'.

Greenland Tundra Reindeer
Rangifer tarandus groenlandicus

But by the beginning of the nineteenth century this range was shrinking apace and the Eastern Elk had already vanished north of the Canadian border. Roads and settlements were spreading, ranches and farms were being established, and the demand for meat and leather increased every year. Added to this was the growing popularity as a sport of running down elk on horseback, an harrassment which the forest and mountain dwelling races further west were spared. But it was the famous 'Elk teeth' which were the greatest cause of their persecution. Indian women had always prized the upper canines as ornaments and many robes may be found in museums where fake 'teeth' can be seen carved out of bone and stitched onto the garments. When the white man became interested in the teeth he would accept no substitutes and, almost incredibly, the fraternal Order of the Elks, which used the teeth as watch-chain insignia, created sufficient demand to wipe out the Eastern Elk. Indeed, they nearly caused the extinction of the other subspecies too – by the time the wanton slaughter was checked the North American wapiti

Eastern Elk (Wapiti) *Cervus canadensis canadensis* Extinct *c.* 1877

had been exterminated from 90 per cent of its former range.

The last pure Eastern Elk on record was shot on 1 September 1877 in Pennsylvania by the famous half-breed elk-hunter, Jim Jacobsen. Six years later six were shot near Elkton in Cavalier County, North Dakota, but these were forest-dwellers and were almost certainly hybrids with the Rocky Mountain Elk (*C. c. nelsoni*) a more massive and paler animal.

MERRIAM'S ELK (WAPITI)

Cervidae
Cervus canadensis merriami
EXTINCT *c.* 1906
New Mexico and Arizona, USA

Merriam's Elk (Wapiti)
Cervus canadensis merriami

South and west of the Eastern Elk's range was the restricted habitat of Merriam's Elk (*Cervus canadensis merriami*) in the mountains of southern New Mexico and Arizona. At the time of its eastern cousin's extinction this spectacular deer was still plentiful as far south as the Mexican border but its decline had begun, largely as the result of the spread of ranching, for the elk could not compete with the Longhorn Cattle on the overgrazed range. Merriam's Elk was, like the Eastern Elk, a reddish animal compared with the Rocky Mountain form but it was the largest of the three subspecies, with a massive skull and more erect antlers.

Vernon Bailey, the authority on New Mexican mammals, believes that it was the antlers of the Merriam Elk which Cortez was shown by Moctezuma; but whether this suggests a previously wider range for the Elk is not clear. In any case, as late as the nineteenth century herds of 2000 or more were reliably reported on the Mexican border. The major decline seems to have started in the 1860s and twenty years later the Elk were gone from New Mexico and restricted to two areas in Arizona: between the Blue and Black rivers in 'beautiful damp meadows in the midst of the dense fir forest on the rolling summit of the Prieto Plateau'; and in the Chiricahua Mountains to the south-east.

Though Merriam's Elk was more at home among timber than the Eastern Elk, its final refuges were not ideal habitat, especially as they were confined to altitudes between 8000 and 10,000 feet, where winters were most severe. Despite legal protection after 1870 this magnificent giant deer was hunted to extinction and the last known herd, little more than a family group, was killed below the peaks of Fly and Chiricahara Mountains in 1906.

BADLANDS (AUDUBON'S) BIGHORN SHEEP

Bovidae
Ovis canadensis auduboni
EXTINCT *c.* 1925
Dakotas and Nebraska, USA

There has been one other North American casualty among the cloven-hoofed family of animals, this time a sheep, the Badlands or Audubon's Bighorn (*Ovis canadensis auduboni*). This extinction is a strange example of an animal's suffering for its ability to survive in the most desolate and hostile environment. Had the Badlands Bighorn been less disposed to take to the cliffs under pressure it might still exist, as its very close relative the Rocky Mountain Bighorn does, in some numbers.

This was no meek or stupid beast. An adult male might weigh 160kg (350lb), was surefooted on apparently sheer cliffs and had a massive armament of horn and muscle which it was quite willing to use on any intruder. It was far from timid and, as the most eastern representative of its family, had taken to grazing on the open prairie, but, instinctively, it kept within reach of the high, isolated 'buttes' or rock outcrops for safety, and this was its downfall. For as the settlers moved in and established their ranches and farms, the buttes became islands and the bighorns became prisoners of their own survival strategy. The further east they had spread from their original mountain homes the more liable were they to this fate, especially at the southern limit of their range, above the Platte River in southern Nebraska.

The total range of this subspecies of bighorn seems to have covered, patchily, North Dakota (including the most forbidding Badlands territory), western South Dakota, Western Nebraska, and probably eastern Montana and Wyoming, though in these areas the distinction between it and its Rocky Mountain cousins became imperceptible.

The Badlands Bighorn was completely cut off from escape routes by the 1880s and it was then only a matter of time before harrassment by dogs and the ability of modern firearms to discount precipices finished it off. The last definite record for North Dakota was the shooting of a gigantic ram on Maggie Creek in the Killdeer Hills in 1905, but its last refuge seems to have been around Harney Peak in the Badlands of South Dakota. It may well have survived there until the 1920s, but the splendid, downcurved horns proved irresistible trophies for hunters and even the Badlands, finally, were inadequate protection for it.

PYRENEAN IBEX

Bovidae
Capra pyrenaica pyrenaica
EXTINCT ? *See text*
French and Spanish Pyrenees

The Pyrenean mountain range between France and Spain contains areas far more forbidding even than the Badlands of the American Dakotas, and here the 'Cabra Montes' or mountain goat, known to us as the Pyrenean Ibex (*Capra pyrenaica pyrenaica*) was in its element. At one time the Pyrenees and the Cantabrian chain were well-stocked

Badlands (Audubon's) Bighorn Sheep *Ovis canadensis auduboni*

Pyrenaean Ibex
Capra pyrenaica pyrenaica

with these agile, spectacular wild goats and the history of the species has been one of slow attrition under pressure of hunting, rather than any sudden or catastrophic decline.

The Pyrenean Ibex's attraction as a trophy was partly the challenge of stalking (and then retrieving) it, and partly its appearance. It stood about 81cm (32in.) at the shoulder, but its horns were up to 102cm (44in.) long and when it had its white winter underfur, its black brow, beard and outer legs stood out in bold contrast. A mounted buck's head not only made a striking decoration, it also conferred upon a hunter the cachet of exceptional skill and hardiness.

According to the foremost authority on the various European ibexes, Angel Cabrera, the population of the Pyrenean Ibex was at its peak about the year 1400, and the slow decline from that date parallels the increased spread and sophistication of firearms. It disappeared first from the French side of the Pyrenees and then from the eastern part of the Cantabrian mountain range, and was unknown in either region by 1850.

By 1907 Cabrera reckoned that there were less than ten of these ibexes in existence, all in Huesca Province round Mount Perdido where he counted 'two old bucks, three females and three or four half-grown individuals'. In 1910 one of his informants, the Count of San Juan, wrote to him: 'I think that probably no more than 10 or 12 Ibexes remain in all the Pyrenean chain' and recounted the pitiful circumstances of one of the last surviving bucks: 'A pair survived recently in the Maladetta – somebody shot the female, and the male sought refuge among a herd of domestic goats and was subsequently killed by the goatherd.' The thought of this most intrepid of cliff-scalers seeking asylum among its degenerate domestic relatives is almost too poignant.

The present status of the Pyrenean Ibex is a vexed issue. There is a group of about 25 animals in the Ordesa National Park, Spain, which some people believe represents the Pyrenean subspecies. (The IUCN, for example, are satisfied that the Pyrenean Ibex survives.) Other authorities are sceptical, to say the least: the fact that other subspecies of Spanish Ibex have long been introduced to this area means that, even if the tiny original herd were indeed Pyrenean Ibexes, hydbidization is very likely to occur.

PORTUGUESE IBEX

Bovidae
Capra pyrenaica lusitanica
EXTINCT *c.* 1892
Galicia and Northwest Iberian Mountains

In 1892 another ibex of the Iberian peninsula had become extinct. The Portuguese Ibex (*Capra pyrenaica lusitanica*) is named as if it were simply a subspecies of the Spanish race, but its appearance was so different that it is hard not to agree with the Portuguese zoologist Carlos Franca when he claims that the 'Cabro do Gerez' was a distinct species.

In size and colouration it was much like the Spanish animal, though inclining towards brown rather than black markings, but its horns were strikingly different from any of the other Iberian wild goats'. They were only half the length of the Pyrenean's (about 51cm, 20in.) but were almost twice as wide and, consequently, much closer together at their bases. The Portuguese Ibex, then, was markedly different both in appearance and balance. But it was no less valued as a trophy and was also hunted by the local mountain villagers for its meat and for the bezoar stones in its stomach which were regarded as potent medicine and antidotes for poisons of all kinds. According to Franca: 'A hunter who would gladly sell the hide, valued the carcass too highly to part with it at any price.'

He described the variety of ingenious traps used to take the *Cabro* and how every part of the animal was utilized. The skins were used as coverlets and the horns both as ornaments and as trumpets or alpine horns to call across the narrow valleys of the northwestern mountains.

Until 1800 the *Cabro* was widespread, its range extending from Borrageiro to Montaglegre, but thereafter its decline was rapid as hunting pressures increased. Unlike the sportsmen, local hunters respected no closed season and were quite happy to take ibexes in May when the kids were smaller and the herds came down to lower altitudes. By 1870 the Portuguese Ibex was a rare animal indeed – the last herd, of about a dozen animals, being recorded in 1886. An old female was taken alive in September 1890 but only survived for three days; two more were found dead the next year, victims of a Galician avalanche; and the last known sighting, near Lombade Pan, was the next year.

Some writers have pointed to other factors than human interference in the ibex's decline. They mention wolf and Golden Eagle predation (surely a constant factor and unlikely to have accelerated); disease from domestic herds; and a disproportionate number of males. This last point seems especially dubious: the bucks were naturally the most hunted and recorded and, besides, the last recorded sightings were all females. There is little doubt that the *Cabro*'s only significant enemy was Man.

Portuguese Ibex *Capra pyrenaica lusitanica*

Quagga *Equus burchelli quagga* Extinct *c.* 1883

The Wild Horses

ZEBRAS, ONAGERS AND TARPAN

QUAGGA

Equidae
Equus burchelli quagga
EXTINCT *c.* 1883
South Africa

Probably no family of animals has had so great a totemistic and economic significance for human civilization as the horse. Our responses to the horse are, and always have been, enormously complex. On the one hand it stands, mythically, for power, grace and, above all, freedom. On the other, it has, through history, carried Man's armies, pulled Man's ploughs and drudged in Man's mines. Then there are the images of the horse as the most obedient and faithful of friends; and as the four-legged embodiment of luck, whose fickle performances at the racetrack have ruined many a good man's life. And the revulsion, almost universal except in the Low Countries, at the thought of eating horse meat involves a curious and contradictory blend of contempt and sentimental reverence.

Partly these responses reflect the horse's own variability. Not only have men developed horses of every conceivable size and shape but, in the wild, horses and asses have adapted to almost every climate and terrain. The Quagga of South Africa (*Equus burchelli quagga*) is probably the best known of all the wild horses, having attracted the dubious glamour that extinction always seems to confer.

When first discovered, however, the Quagga was simply regarded as one zebra among many and was often thought to be the female of the Burchell Zebra. The Hottentots, who gave the name 'Quahah' in imitation of the animal's shrill, neighing cry, gave the Burchell the same name. So it would be inaccurate to claim that the Quagga was particularly singled out for hunting or exploitation. Far more significant to its extinction was its geographic restriction to the old Cape Colony veldt and, in lesser numbers, to the Orange Free State. In the Cape it was, in fact, the only zebra on the plains. Thus the colonizing Boers found it the most obvious source of food for their native servants and of hides for domestic use and export; and they shot Quaggas in their thousands. As the Boers moved north they exploited the more numerous species of zebra in the same way but with less drastic consequences.

The Quagga was basically a brown, rather than a striped, zebra with white legs and tail. It had no distinct markings on the hind quarters and only vague mottlings on its back. So its only striped parts were its head and neck. In conformation it was far more horse-like than any of the other zebras which, with the exception of the Burchells, are essentially large-headed, donkey-like creatures.

The Quagga's mane was also distinctive, being described by an early observer as 'curious, appearing as if trimmed by art'. It was an immensely energetic and highly-strung animal, the stallions being given to occasional fits of rage. In fact the London Zoo's one chance of breeding Quaggas in the 1860s was foiled when the stallion beat itself to death against the wall of its enclosure.

In the early days of settlement, the Boers of Cape Colony kept tame Quaggas as guards to their domestic stock at night, knowing that their 'watchdogs' would not only raise the alarm but very probably attack any intruder – man or beast – viciously. It is, perhaps, strange that such animals could be tamed, but in England in the 1830s there was a vogue for Quaggas as harness animals and Sheriff Parkins could be seen around London seated behind a pair of the exotic imports. They were said to have much better 'mouths' for harness than Burchell's Zebras which were also fashionable between the shafts for a time. Whether the harness Quaggas were gelded or not is not documented, but it would seem likely.

In its natural state the Quagga formed what has been called a 'triple alliance', typical of the southern zebras, for defence against predators. It was nearly always to be found in the company of Wildebeest (White-tailed Gnus) or hartebeest, and ostriches. This has been explained, hypothetically, as a synthesis of varied talents: the birds' eyesight, the antelopes' powers of smell, and the Quaggas' acute hearing. Certainly such a group of animals, grazing as they did on the open plains, would have been well protected against surprise by any natural predators and their chief enemy, the lion, probably caught very few healthy adult Quaggas.

But the Boers, of course, had horses, firearms and, for live captures, a form of lariat. They found the great herds of Quaggas and antelope easy pickings indeed and, both in the Cape and north by the Orange River, were reported to be 'as much interested in the hide business as in their general occupation of farming'. For their own use, sacks for storage and transportation were normally made from the sturdy, lightweight Quagga skins and were still to be seen in everyday use long after the herds themselves had disappeared and the shrill warning cries 'kwa-ha-ha, kwa-ha-ha, quickly repeated' were only a memory preserved in the animals' name.

The destruction of the great herds which abounded in the 1840s seems to have taken about 30 years. The last wild Quagga was killed in 1878, and in 1883 the last captive Quagga (a female) in the Amsterdam Zoo died, rendering the species extinct.

BURCHELL'S ZEBRA

Equidae
Equus burchelli burchelli
EXTINCT *c.* 1910
South Africa

Recurrent rumours of Quaggas reappearing in the last hundred years usually refer to sightings of Burchell's Zebras with aberrant markings, and it is quite possible that there was inter-breeding between Quagga and Burchell in the past and that occasional 'throw back' foals are born. But the nominate race of Burchell's Zebra (*Equus burchelli burchelli*), the white-legged, white-tailed form which most resembled the Quagga, has itself been extinct since the beginning of the century. The last known specimen died in London Zoo in 1910, having been a captive almost all its life.

This zebra was restricted to the Orange Free State and southern Bechuanaland and its range partially overlapped that of the Quagga, which, as we have seen, was sometimes regarded as a female Burchell. The essential differences were the stripes on the Burchell Zebra's body and haunches and the curious 'shadow stripes' between its black markings. Though the Burchell Zebra's mane was erect it was more ass-like than the Quagga's. To the extent that the Voertrekkers distinguished between the two animals, they called the Burchell 'Bontequagga'. It abounded 'in countless thousands' on the plains,

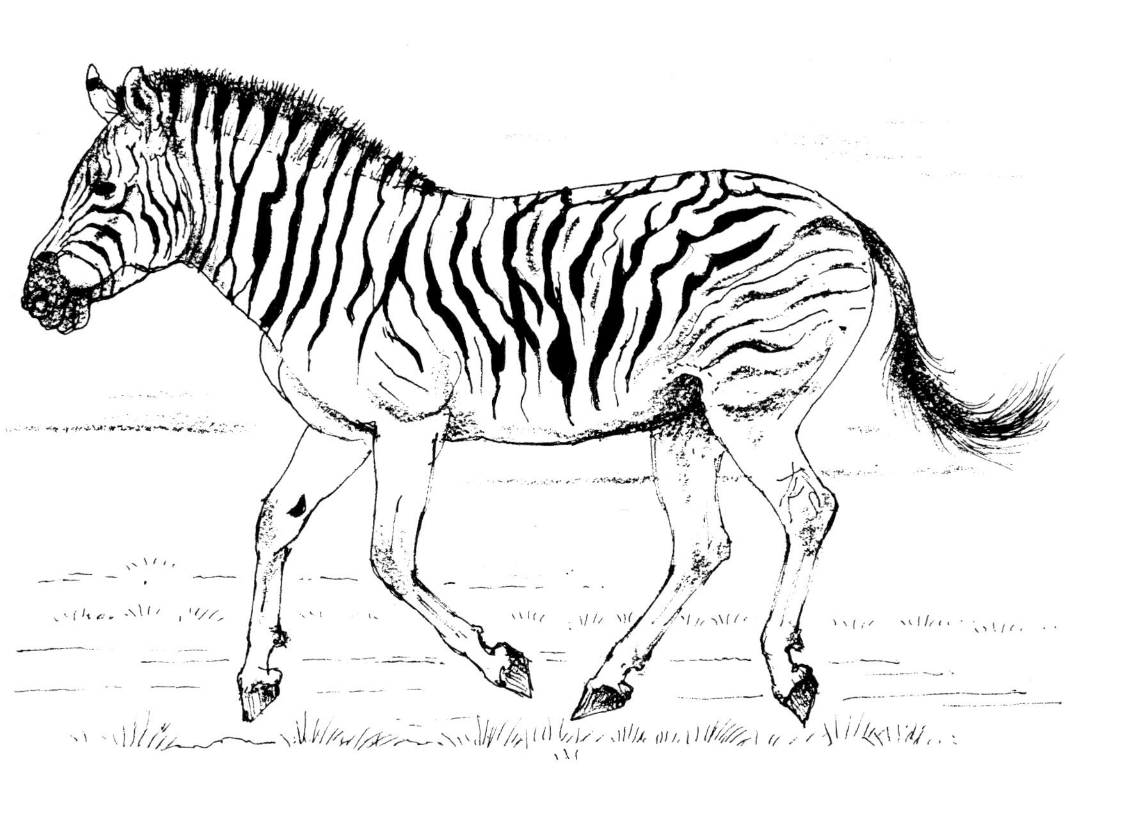

Burchell's Zebra *Equus burchelli burchelli*

grazing like the Quagga in the company of ostriches and wildebeest (in this case the Brindled Gnu), but it had become extremely rare by 1850, as both hunters and farmers took their toll. Only the survival of three subspecies to the northwest and northeast makes it loss less of a tragedy than the utter extinction of its more elegant cousin, the Quagga.

SYRIAN ONAGER

Equidae
Equus hemionus hemippus
EXTINCT *c.* 1930
Middle East

The swiftest, hardiest and smallest members of the horse family in historical times have undoubtedly been the wild asses of North Africa, Asia Minor, Persia and Tibet. So much do they give the lie to the common conception of the ass or donkey as a stolid, slow creature that the biblical name of Onager is to be preferred. In medieval times subspecies of this genus became extinct both in North Africa and Anatolia, but the only modern casualty has been the Syrian Onager (*Equus hemionus hemippus*) which was last seen in the wild in 1927, and in captivity around 1930.

This onager's range was far wider than Syria, for it occurred in great numbers in Palestine, Arabia and the whole of Mesopotamia within the last 100 years. Its speed and its ability to withstand the worst conditions of the Hammad and Nafud deserts left its numbers unaffected by over 2000 years of intensive hunting by successive cultures in the area. It was only the coming of firearms and automobiles which turned the scale against it.

From the earliest times this tiny (90cm, 3ft at withers) horse was regarded as game and food rather than as a potential beast of burden. Bas-reliefs uncovered at the capital of ancient Assyria, Nineveh, depict the hunting expeditions of King Ashurbanipal around 650 BC, and one slab in particular shows two of the king's servants lassoing an onager. They must have been especially skilful and lucky huntsmen: even this boastful carving shows the rest of the asses escaping and outdistancing their pursuers with ease.

The humbler inhabitants of the region were less chivalrous: they were hunting for the pot and concentrated on taking the young onagers in the spring foaling season. The Italian traveller Della Valle described a captive 'wild ass or little onager' in Basra, southern Iraq, in 1625 and the English excavator of Nineveh, Sir Austen Layard, reported 280 years later that the Bedouin 'bring the foals up with milk in their tents . . . They are of a light fawn colour – almost pink. The Arabs still eat their flesh.'

In 1850 the onager was becoming scarce in the Syrian desert (Badietesh Sham) and in Palestine but, according to the Englishman Canon Tristram, was still common in Mesopotamia and could be seen in the summer travelling in great white herds as far as the Armenian mountains.

The first real threat to the onagers' survival came with World War I when, with the Arab push towards Damascus, the whole area was overrun with heavily armed Turks, Bedouin and British troops, and the automobile began to replace camel and train in opening up the deserts. According to the German zoologist, T. Aharoni: 'The movements of Bedouin troops during the Great War and the more recent incursions of the Wahabi tribes . . . have pushed back these extraordinarily shy, freedom-loving creatures into the heart of the desert. They appear so sporadically now that most Bedouin tribes

Syrian Onager *Equus hemionus hemippus* Extinct *c.* 1930

have not seen them at all in recent years.'

Aharoni was writing in 1930 but there is every likelihood that the onager was already extinct by that time, though the custom aaong such tribes as the Shaleib of releasing their domestic ass mares to be impregnated by the onagers meant that the blood line continued in diluted form. As far as records show, however, the last wild Syrian Onager was shot in 1927 as it came down for water at the Al Ghams oasis not far from Lake Azrak in the Sirhan depression of north Arabia. This lava-bed district seems to have been one of the last three pockets of survival for the wild asses. The other, similar, areas were the Jebel el Druz in Southern Syria and the Jebel el Sinjar on the Iraq-Syria border.

It was from this last region that the Schonbrunn Zoo received a specimen which was still alive as late as 1928. It may have been the last pure-bred Syrian Onager in the world, although some writers continue to express the pious hope that, with the Arabian Ostrich which often kept them company, some onagers may still be hiding out in the desert fastnesses of Saudi Arabia or Oman. It seems probable, though, that the definitive work on the distribution of recent Equidae, written by Otto Antonius in 1937, contains the true epitaph of the Syrian Wild Ass: 'It could not resist the power of modern guns in the hands of the nomads, and its speed, great as it may have been, was not sufficient always to escape from the velocity of the modern motor car which more and more is replacing the Old Testament Camel-Caravan.'

TARPAN

Equidae
Equus ferus
EXTINCT *c.* 1887
Eastern Europe and Western Asia

The zebras and wild asses, graceful and energetic though they are, bear little resemblance to our modern ideal of equine elegance as exemplified in Arabs, Thoroughbreds and other 'warm-blooded' breeds. The ancestor of all these 'warm-bloods', a plateau-dwelling pony called the Tarpan (*Equus ferus*), became extinct within the last hundred years; its blood still flows in herds of mixed-blood feral horses; and some controversial 'reconstructed' specimens may be seen today in zoos.

The Tarpan differed from the primeval ponies of the Forests (ancestors of today's draught-horses) and the Steppes (the surviving Przewalski Horse) in its light build, slender limbs, small concave head, small ears, large eyes and abundant tail and mane. It was alert, intelligent and graceful – exhibiting, in a crude and diminutive form (it only stood 13 hands, 132cm), all the qualities that are valued today in Arab and Thoroughbred horses.

It was predominantly mouse-coloured, with some tan variants, and its flowing mane and tail were black. In winter (and the Asiatic steppes, the Ukraine and the Gobi Desert were particularly harsh parts of its early range) its hair became long and soft. Reports of white Tarpans seem reliable too, and probably account for the near-mythical white horses uncovered by ivory hunters in the permafrost of Siberia and still retained in the tribal memory of the Lamuts. In recent times the Tarpan was restricted to eastern Europe. It died out in the Ukraine around 1850 and the last known pure-bred Tarpan was killed near Askamia Nova in Russia on Christmas Day in 1879. Eight years later the last captive Tarpan died in Poland and the government set out, too late, to 'save' it.

They collected from peasant farmers ponies which had obvious Tarpan characteristics, and established herds of them in such forest reserves as Popielno. These herds certainly approximate in apapearance and lifestyle the original Tarpan, but the process of 're-construction' was taken even further by Munich Zoo experts. They bred selectively from animals of the Popielno and Bialowlieza herds and (checking skeletal structures at every generation) claimed to have recovered the Tarpan in its pure form. These are the 'Tarpans' which may be seen in the larger modern zoos.

The reasons for the decline of the wild Tarpan are complex. From the earliest times it was hunted for food and, because of its spirited independence, was usually cliff-driven or otherwise massacred: for at close quarters it was more than a match for any domesticated horse. This skill in combat (Tarpan stallions frequently killed rival horses almost twice their weight) led in turn to a dilution of the wild blood as the Tarpans invaded or hived off herds of domestic mares. In later times, of course, the wild ponies had to face invasion and settlement of their range and the spread of firearms.

A Mongolian legend reveals the admiration which early men felt for the Tarpan and the superiority it was accorded over the other Asiatic pony, the Przewalski. The legend deals with the Torguls, a centaur-like tribe descended from Tarpan the stallion and the beautiful princess Irgit. At the birth feast of their son Torgut, the assembled horses do battle with the wolves and destroy 20,000 of them. The supposedly deathless Tarpan then gallops off to the east bearing the young prince on his back. Sadly, they were riding to extinction.

Tarpan *Equus ferus*

Steller's Sea Cows *Hydrodamalis gigas* Extinct *c.* 1767

The Sea Mammals

SEA COWS, SEALS, SEA MINK AND WHALES

STELLER'S SEA COW
Dugongidae *Hydrodamalis gigas* EXTINCT *c.* 1767 Bering Sea

'Their capture was effected by a large iron hook, the point of which somewhat resembled the fluke of an anchor, the other end being fastened by means of an iron ring to a very long and stout rope, held by thirty men on shore. A strong sailor took this hook and with four or five other men stepped into the boat, and one of them taking the rudder, the other three or four rowing, they quietly hurried towards the herd. The harpooner stood in the bow of the boat with the hook in his hand and struck as soon as he was near enough to do so, whereupon the men on shore, grasping the other end of the rope, pulled the desperately resisting animal laboriously towards them. Those in the boat, however, made the animal fast by means of another rope and wore it out with continual blows, until tired and completely motionless, it was attacked with bayonets, knives and other weapons and pulled up on land. Immense slices were cut from the still living animal, but all it did was shake its tail furiously and make such resistance with its forelimbs that big strips of the cuticle were torn off. In addition it breathed heavily, as if sighing. From the wounds in the back the blood spurted upward like a fountain. As long as the

head was under water no blood flowed, but as soon as it raised the head up to breathe the blood gushed forth anew. . . . The old and very large animals were much more easily captured than the calves, because the latter moved about much more vigorously, and were likely to escape, even if the hook remained unbroken, by its tearing through the skin, which happened more than once.'

This might have been a crew of seamen slaughtering a trapped herd of whales, but it was not. It was the hunting of another true giant of the oceans, that strange creature called Steller's Sea Cow (*Hydrodamalis gigas*), in the arctic waters of the Bering Sea.

Measuring 6–9m (20–30ft) in length and weighing up to 6,400kg (14,000lb), these animals totally dwarfed all other sea mammals outside the cetacean group. On land they were only rivalled in size by the elephants. But, size apart, the name 'Sea Elephant' would in fact be far more appropriate than 'Sea Cow'.

Steller's Sea Cow belongs to the order Sirenidae, which comprises only three genera: Steller's Sea Cow in the Pacific, and the far smaller Manatees of the Atlantic coasts (*Trichechus manatus*) and the Amazon (*T. inunguis*); and the Dugongs (*Dugong dugon*) of estuaries on the Indian Ocean. Every surviving species is endangered, especially in their freshwater haunts. Though outwardly appearing to be an order somewhere between the whales and the seals, these creatures are unrelated to either. They were descended from land animals ancestral to the Proboscideans, the order which is today represented only by the elephants. Their adaptation to water resulted in a whale- or seal-like shape.

The description of the sea cow hunt quoted above was made by the first European ever to see these creatures. This was Georg Wilhelm Steller, the naturalist and physician attached to the last ill-fated Russian exploratory expedition led by the Dane, Vitus Bering. The party was virtually wrecked on what is now Bering Island in 1741. At that time the Sea Cows numbered thousands in the shallow waters about Bering Island and nearby Copper Island. Incredibly, through terrible and wasteful slaughter, this 'elephant of the sea' became extinct not more than 27 years after its discovery.

Steller was the only scientist to observe this animal, and his descriptions, along with a few skeletons, are virtually the only proof extant of the animal ever having existed. Fortunately (if such a word can be applied to such tragic circumstances) Steller recorded the Sea Cow's nature, habits and biology in studious detail. Furthermore, he wrote of the animals with compassion, and in their slaughtering could not but help applying human attributes to their behaviour:

'When an animal caught with the hook began to move about somewhat violently, those nearest in the herd began to stir also and feel the urge to bring succour. To this end some of them tried to upset the boat with their backs, while others pressed down the rope and endeavoured to break it, or strove to remove the hook from the wound in the back by blows of their tail, in which they actually succeeded several times. It is most remarkable proof of their conjugal affection that the male, after having tried with all his might, although in vain, to free the female caught by the hook, and in spite of the beating we gave him, nevertheless followed her to the shore, and that several times, even after she was dead, he shot unexpectedly up to her like a speeding arrow. Early next morning, when we came to cut up the meat and bring it to the dugout, we found the male again standing by the female, and the same I observed once more on the third day when I went there by myself for the sole purpose of examining the intestines.'

In the face of the extraordinarily harsh and desperate circumstances of his castaway residence on Bering Island, Steller showed considerable sensitivity in describing the sea beasts' treatment of their young, and their courtship. He wrote: 'These animals love shallow and sandy places along the seashore, but they spend their time more particularly about the mouths of the gullies and brooks, the rushing fresh water of which always attracts them in herds. They keep the half-grown and young in front of them when pasturing, and are very careful to guard them in the rear and on the sides when travelling, always keeping them in the middle of the herd. With the rising tide they come in so close to the shore that not only did I on many occasions prod them with a pole or a spear, but sometimes even stroked their back with my hand. . . . In the spring they mate like human beings, particularly towards evening when the sea is calm. Before they come together many amorous preludes take place. The female, constantly followed by the male, swims leisurely to and fro eluding him with many gyrations and meanderings, until, impatient of further delay, she turns on her back as if exhausted and coerced, whereupon the male, rushing violently upon her, pays the tribute of his passion, and both give themselves over in mutual embrace.'

When Steller first saw these beasts, he did not know what to make of them because he could not see their true shape as they swam, grazing on seaweed, in the bays around the island. All he could see were the massive blackish backs like so many overturned boats moving slowly about, and every few minutes a snout emerging in front with a snorting breath. It was only later, when he saw a beached animal, that he realized it was a sirenian.

Beyond its unbelievable size, the Steller Sea Cow differed from all other sirenians in its extraordinarily thick outer skin which looked like 'the bark of an old oak', and in the strange horny plates of its gums which served to cut and grind the vegetable diet of these toothless mammals. Like all mammals of this order, the Sea Cow had no back legs or even a trace of a pelvic bond, but typically had rather short (67cm, 26½in.) stump-like forelimbs. However, these were unique appendages. The outer skin at the ends of these limbs was much thicker and harder than elsewhere, so that the ends of the feet were rather like the hooves of a horse, only more pointed and thus more suited for digging. These hooves were smooth and rounded above, flat and slightly hollowed below, where they were thickly covered with 1.25cm (½in.) bristles like a brush. With these seemingly short and awkward limbs, the great beasts swam, crawled along the bottom, held onto slippery rocks, caressed their mates, and tore seaweed from the sea pastures like horses paw the ground with their hooves.

Steller wrote: 'All they do while feeding is to lift the nostrils every four or five minutes out of the water, blowing out air and a little water with a noise like that of a horse snorting. While browsing they move slowly forward, one foot after the other, and in this manner half swim, half walk like cattle or sheep grazing. Half the body is always out of the water. Gulls are in the habit of sitting on the backs of the feeding animals feasting on the vermin infesting the skin, as crows are wont to do on the lice of hogs and sheep. They do not eat all kinds of seaweed promiscuously, but select. . . . Where they have been staying even for a single day there may be seen immense heaps of roots and stems. Some of them when their bellies are full, go to sleep lying on their backs, first moving some distance away from shore so as not to be left on dry land by the outgoing tide.'

The meat of the Sea Cow, which was redder than land mammals' meat generally, was evidently delicious. The adult animals' flesh was said to be superior to beef, and when properly prepared made an acceptable substitute for corned beef. The younger animals' meat was said to be very like veal, and the flesh of both adult and young would swell to nearly twice its size when cooked. There was also a 10cm (4in.) thick fat layer under its 2.5cm (1in.) thick outer skin. This fat was compared to 'May butter' for richness, its fine flavour was like that of sweet almond oil, and it burnt with a clear, inodorous and smokeless flame.

Such a ready food source in so relatively barren an arctic region was soon exploited by the many others who followed Bering's expedition to hunt for furs on the islands and elsewhere. The sea cows were used to provision these ships and feed their crews through the long winters. Apart from the animals' appealing taste and the fact that one sea cow could feed 33 men for one month, both meat and fat were attractive to the shipmasters because they kept well without putrifying even in very hot weather.

Unrestrained hunting resulted in shamefully wasteful slaughter. Because of the beasts' sheer size, it is estimated that four animals were critically wounded and abandoned for every one that was successfully beached and killed for food. In 1754, a Russian mining engineer named Jakovleff who was looking for copper on Copper Island, reported on the disastrous hunting methods which that year had wiped out the last of the beasts there. He petitioned the Kamchatkan authorities to take some measures for the protection of the animals on Bering Island, for he saw no purpose in so wastefully destroying the only major food source available to men in this region. No measures of consequence were taken and in 1767 the last known specimen of Steller's Sea Cow was slaughtered on Bering Island.

CARIBBEAN MONK SEAL

Phocidae
Monachus tropicalis
EXTINCT *c.* 1952
Caribbean Sea

The first New World animal to be logged in the journals of Christopher Columbus on his historic voyages of discovery, was the Caribbean Monk Seal (*Monachus tropicalis*). His crewmen slaughtered eight of them on the islet of Alta Vela to the south of Hispaniola. These large seals, 240cm (8ft) long and 180kg (400lb) in weight, were once abundant throughout the Caribbean and the pirate naturalist, William Dampier, wrote that in the late seventeenth century, the Caribbean Monk Seal was the source of a profitable oil industry. This exploitation and the associated fur trade expanded to an extraordinary degree in the eighteenth century.

The monk seals form a distinctive sub-family (Monachinae) of seals which consists of three widely separated species living in sub-tropical waters. The other two are the Mediterranean Monk Seal (*Monachus*

Caribbean Monk Seal *Monachus tropicalis* Extinct *c.* 1952

monachus) and the Hawaiian Monk Seal (*Monachus schauinlandis*). As with nearly all fur-bearing sea mammals, the monk seals were hunted on a massive scale through the eighteenth and nineteenth centuries. The Mediterranean species today has a population of less than 500 with virtually no hope for recovery or even survival beyond the end of the century. The Hawaiian species, which was similarly reduced to around 500 by 1951, has been more fortunate in that effective protection measures could be enforced – and has modestly increased to over 2000. For the Caribbean Monk Seal the critical point had been passed by the end of the nineteenth century, and it has become extinct within the last 30 years.

The last known populations of Caribbean Monk Seals were on the Triangle Keys, a series of small sandy islets off the Yucatan Peninsula. In 1911, fishermen came and slaughtered all of the 200 remaining seals in this region. This was the virtual end of the species.

Although a few seals have been observed since then, the last reliable sighting was in Jamaican waters in 1952. A thorough aerial survey of all possible habitats in 1972 and an expedition in April 1980 failed to find any trace of the Caribbean Monk Seal. Both searches concluded that the animal was undoubtedly extinct.

SEA MINK

Mustelidae
Mustela macrodon
EXTINCT *c.* 1880
New England

The little-known Sea Mink (*Mustela macrodon*) of New England and the Canadian maritime provinces, although not properly a sea mammal, was restricted to coastal waters and suffered from fur hunters quite as much as the seals and otters.

For a mink, this creature was a giant – easily the largest in the world. At the turn of the century Manly Hardy wrote in *Forest and Stream* magazine that his family, who were fur-buyers in Maine, recognized this very large mink as a distinct form which was commonly available until 1860. The animal is known to have bred as far north as Nova Scotia where, in 1867, a J. B. Gilpin recorded a Sea Mink measuring 82.6cm (32½in.) in overall length, whereas the largest Common Mink trapped in New England did not exceed 57.6cm (23in.) and that was an oversized male. The Sea Mink was not just longer, but larger and fatter, giving it a dressed pelt about twice the size of the smaller inland species'. The pelt evidently had a totally different odour from the smaller species', and the fur was coarser and redder.

The Sea Mink's remains are commonly found in Indian shell-heaps on the coast and islands of Maine. Presumably the Indians of the region hunted the animal for its pelt, and this may have contributed to its decline. However, it was undoubtedly the highly competitive European fur trade that dealt the final extinguishing blow. It was the fur-buyers who gave it the name 'Sea Mink', and its size would have guaranteed it as a preferred species.

In 1880, a Sea Mink was sold to a fur-buyer at Jonesport, Maine. It had been killed on one of the nearby islands and measured over 66cm (26in.) without the tail. It was the last record of the species.

The most celebrated of sea mammals are, of course, the whales. Quite properly, one of the major causes of environmentalist politics is the protection of whales. Exclusive of the dolphin and porpoise groupings (and the two 'pygmy whale' species) there are only ten species of Great Whales, five of which have become 'commercially extinct' during this century. However, no recorded biologically distinct species or subspecies has been totally exterminated yet – although it is almost certain that, had proper examination been made of the Grey Whale (*Eschrichtius robustus*) in the Atlantic before its extirpation by the New England whalers in 1730, the Atlantic Grey Whale would have been named a separate subspecies from the surviving (though endangered) Pacific Grey Whale.

The titanic Blue Whale (*Balaenoptera musculus*) which may measure 30.5m (100ft) and weigh 157.5 tonnes (160 tons), is the largest of all whales. It is also among the rarest, with a world population of about one per cent of the quarter million that inhabited the oceans a century ago. Similarly, the Bowhead (*Balaena mysticetus*) and the Black Right Whale (*Balaena glacialis*) each have a population of less than 1000. The other two commercially extinct whales, the Grey Whale and the Humpback (*Megaptera novaeangliae*) survive with less than ten per cent of their original populations. For each of these species, it is the razor's edge. It may well be that some, even with total protection, have been so depleted that biological extinction will overtake them without further persecution.

The campaign to save the whales is, of course, an effort of symbolic as well as intrinsic value. In real terms, whaling is of little or no importance to the economy of any modern nation. If the countries of the world cannot come together and successfully overrule such a small economic pressure group as the whaling industry, in order to permit the survival of the largest animals ever to inhabit the planet, there can be no real hope for the hundreds of other endangered species on the brink of extinction.

Sea Mink *Mustela macrodon*

Thylacines or Tasmanian Wolves *Thylacinus cynocephalus* Extinct *c.* 1933

The Marsupials

THE THYLACINE, KANGAROOS AND BANDICOOTS

THYLACINE OR TASMANIAN WOLF
Thylacinidae *Thylacinus cynocephalus* EXTINCT *c.* 1933 Tasmania

For eighteenth and nineteenth-century Europeans, Australasia was as much a new world as the Americas, and if its tribal people were less numerous and varied, its fauna (especially the quadrupeds) was exotic almost beyond belief. The new animal kingdom of the marsupials fed the Old World's appetite for marvels and created such disparate mythologies as Tigers and Devils from Tasmania and Kanga and Roo from Pooh Corner.

In fact the marsupials do connect the two New Worlds, for they seem to have developed in South America and spread from there, and species survive in the western hemisphere – notably the Virginia Oppossum. Marsupials are principally restricted to Australasia because that region became isolated before the spread of more evolved mammals which seem in the majority of cases to oust their more primitive, pouched cousins.

Thus the story of the marsupials since European settlement has been one of retreat and decline, and we can only guess at the effect of the earlier Aboriginal introduction of the dingo, or feral dog. We do know that for at least a thousand years the marsupial equivalent

of the dingo, the Thylacine or Pouched Wolf, had been extinct on mainland Australia and restricted to dingo-free Tasmania.

The carnivorous marsupial called the Thylacine or Tasmanian Pouched Wolf (*Thylacinus cynocephalus*) was undoubtedly the strangest 'wolf' the world has ever seen. One early European observer referred to it as 'a kangaroo masquerading as a wolf', and it was also known as the Tasmanian Tiger, the Kangaroo Wolf, the Zebra Wolf and, occasionally, the Hyaena Oppossum. It was the only member of its genus, and of its family (*Thylacinidae*). It had the head and teeth of a wolf, the stripes of a tiger, the tail of a kangaroo and the backward-opening pouch of an oppossum. It measured 150cm to 180cm (5–6ft) in overall length, but had neither the weight nor speed of the true wolf.

The Thylacine fed originally on kangaroos, wallabies and ground birds. It always hunted singly at dusk, night or dawn, following its prey by scent at a leisurely trot as it was incapable of any great speed. If pressed, it broke into a 'shambling canter', and could rise on its hind legs and hop if it encountered difficult obstacles. A Thylacine never chased quarry flat out, but doggedly pursued it until it showed signs of exhaustion and then rushed in.

Adults made lairs in caves, among rock piles, in hollow logs or in hollow trees where they spent nearly all the daylight hours. Usually the female had four cubs which it carried in its pouch for three months and then left in a kind of 'nest'. In 1920, a Mr Flinty of Smithton was barred by a female from crossing a creek on a fallen tree. Once across, he 'searched the bushes and found four young secreted in a dry fernbed under the drooping and still attached dead fronds of a treefern. These reached the ground all round the butt, thus forming a natural tent-like shelter and a perfect camouflage.'

Although the Pouched Wolf would not have matched the true wolf as a hunter, it was armed with jaws of great strength whose gape was far greater even than that of the largest Grey Wolf. The Thylacine invariably killed its prey by crushing the skull.

When men attempted to hunt the Tasmanian Wolf with dogs, it was often at cost, for it had no fear of even the biggest kangaroo hounds. One Ronald Gunn reported that even a pack of dogs would refuse to move in on an old male Thylacine once they got it at bay. According to one hunter, H. S. Mackay: 'A bull terrier once set upon a Wolf and bailed it up in a niche in some rocks. There the Wolf stood with its back to the wall, turning its head from side to side, checking the terrier as it tried to butt in from alternate and opposite directions. Finally the dog came in close and the Wolf gave one sharp, fox-like bite, tearing a piece of the dog's skull clean off, and it fell with the brain protruding, dead.'

Despite this murderous power, it seemed universally accepted among those with experience of the Thylacine that it never attacked humans unless it was trapped or at bay. However, there were odd incidents of old and half-blind Thylacines making weak and almost ludicrous attacks on settlers. Of those recorded, all failed and the pathetic animals were found to be starving and almost toothless, being easily killed or driven off with sticks or, in one case, a poker swung by a child.

Nevertheless, the Thylacine suffered the same persecution as every creature which has been named, however arbitrarily, a wolf. Indeed, it seems the Thylacine was a chief bogey of Tasmania to the point of 'superstitious dread'. The Europeans killed them at every opportunity and often smashed them to a pulp once dead, so that the carcasses were unavailable even for scientists. They gained an unreasonable reputation as sheep-killers, for even the statistics of the day – totally biased against the Thylacine – showed far more sheep killed by dogs. Furthermore, old European beliefs about wolves persisted among the settlers who often insisted – totally falsely – that the Tasmanian Wolf killed for blood, and like a vampire sucked all its victim's blood from the jugular.

In 1888 the Tasmanian government offered a bounty for the Thylacine, although, since 1840, the Van Dieman's Land Company had set their own bounty on the animals. Between 1888–1914 at least 2,268 Thylacines are *known* to have been killed and turned in. In 1910 an epidemic rather like distemper, and possibly brought by domestic dogs, may have reduced the already dangerously low population.

In 1936, the government totally reversed its stance and granted the Thylacine complete protection, imposing severe fines on anyone killing one. The gesture was almost absurd. The last authenticated killing of a wild Thylacine was in Mawbanna in 1930; the last one captured was in 1933. This was kept in Hobart Zoo but died the same year. There have been later claims of Thylacine sightings and even killings. One 'kill' produced only hair and blood samples of debatable origin; and another, a brindled greyhound. Since 1933 there has been no acceptable evidence of the animal's existence. At least two major expeditions have been launched to search out the Thylacine since then, with no result.

Irony seems to be a thing that totally eludes governments: in 1966 the Tasmanian officials declared a huge game reserve in the south-west, extending from Low Rocky Cape to Kellista to South West Cape, where cats, dogs and guns were prohibited. This reserve was assigned for the preservation of the Thylacine, which was last seen 33 years before, and which until then this same government had actively encouraged its citizens to exterminate.

TOOLACHE WALLABY

Macropodidae
Wallabia greyi
EXTINCT *c.* 1940
South Australia

It should be said that Australian birds and mammals are almost notorious for reappearing years after being declared extinct, and becoming 'lost' again once assumed to be safe. Australia is still a continent of vast space and wilderness and one can hope, in some cases, that the extinctions recorded here just might be, like reports of Mark Twain's death, 'somewhat exaggerated'. There is no room for doubt, however, about the utter disappearance of the most beautiful of all the Kangaroo family, the Toolache or Grey's Wallaby (*Wallabia greyi*), which died out about 1940.

Inhabiting the open plains of South Australia the Toolache – with maximum measurements of 84cm (32in.), body, and 73cm (28½in.), tail – was highly valued both as a fur-bearer and as a game species and is thus well-documented by contemporaries. We know from John Gould, that most meticulous nineteenth-century illustrator of Australasian animals, that the Toolache favoured open ground, 'intersected by extensive salt lagoons and bordered by pine-ridges', and its dependance on such habitat may have been one weakness in its capacity to survive change. It was, though, wonderfully adapted to its environment: 'I never saw anything so swift of foot as this species,' Gould declared '. . . it does not appear to hurry itself until the dogs have got pretty close, when it bounds away like an antelope, with first a short jump, and then a long one, leaving the dogs far behind it.' Professor F. Wood Jones, who led a fruitless last-ditch effort to save the Toolache in the 1920s, called it 'by far the fleetest of all the Wallabies' but emphasized the importance of its valuable fur to its decline: 'the beautiful pelts have been marketed in very large numbers in the salesrooms of Melbourne'. Between Gould's observations (1852) and Jones' (1924) the Toolache population had dwindled from 'swarms' to 'five or six individuals'.

One must remember that, hunting and fur-gathering apart, there was in the early part of this century a bonus of sixpence paid on *all marsupial scalps* so that the temptation for poor or young Australians to shoot open-country wallabies must have been almost irresistible. An added pressure was the spectacular spread of imported foxes which preyed extensively on the Toolache 'joeys' whose only previous enemy had been the Wedge-tailed Eagles. Many authorities lay the most blame on the foxes, though it seems unlikely that they could directly have harmed the swift adults except, perhaps, by night.

Toolaches were still seen commonly around 1910: solitary or in small groups where the land was arid, and in true herds where grass was lush and abundant. But by 1923 the only group of these Wallabies surviving in their natural setting was a band of about 14 on the Konetta sheep run of Mr J. Brown near the town of Robe. In May of that year Professor Wood Jones's campaign to save the Toolache bore fruit, but with disastrous consequences. A number of the Wallabies were run down and captured to be transferred to a sanctuary on Kangaroo Island, but the spirited creatures literally ran themselves to death trying to escape their well-intentioned pursuers. Another drive the next year had the same result.

The situation was desperate now and another effect of Wood Jones's publicity was to make the Toolache seem a desirable rarity for trophy hunters. Some of the Konetta herd were shot and individuals lurking in the open 'stringy-bark' country were hunted out with increased fervour.

In 1937 a Toolache doe was rescued from the jaws of two kangaroo dogs at Konetta by Mr Brown. Though badly mauled, the Wallaby survived and lived for another couple of years, probably the last wild representative of her kind. The last Toolaches on Earth died in the Adelaide Zoo, without having bred, a few years afterwards.

EASTERN HARE-WALLABY

Macropodidae
Lagorchestes leporides
EXTINCT *c.* 1890
Southeastern Australia

The early settlers of southeastern Australia found a great variety of Kangaroo-like marsupials on the interior plains, varying in size from great 'boomers' which could turn on a hunter and disembowel him with a stroke of the hindfoot, to creatures smaller than a rabbit. Next to the unfortunate Toolache, the Hare-Wallabies (*Lagorchestes*), were probably the most favoured game animals in these districts. The colonists gave them this name because they were so reminiscent in their behaviour of the Brown and Belgian Hares of Britain and Western Europe. Lying up in 'forms' among the spinnifex tussocks or

Toolache Wallaby *Wallabia greyi* Extinct *c.* 1940

under bushes, the Hare Wallabies would bolt, when flushed, and show an astonishing if short-lived burst of speed. They could be shot, snared, netted or 'coursed' with hounds and were the closest thing available to the sport, or more often poaching, that the New Australians had enjoyed in the Old Country.

There were four or five loosely-related species of Hare-Wallaby, differing somewhat in appearance but mostly in social behaviour. Some were colonial, one (*L. asomatus*) is known only from a single skull found in 1931 and is a complete mystery, while the Eastern and Western Hare-Wallabies were solitary browsers on the open grasslands. The Eastern form (*Lagorchestes leporides*) has not been seen since 1890 and is presumed to be extinct.

Like all this group the Eastern Hare-Wallaby had a curious ring of bright orange hair around each eye; with the added peculiarity of black 'elbow-patches' which showed up distinctly against its brown fur when it ran. Like the Toolache, it was renowned for its speed, but its leaping abilities – if we can believe contemporary accounts – were still more phenomenal. John Gould, a normally cautious authority, describes two of his fastest dogs coursing one for .8km ($\frac{1}{2}$ mile) and heading it right towards him. When the Hare-Wallaby was within 6m (20ft) of him it 'bounded cleanly' over his head.

The introduced European Fox proved to be the most deadly enemy of this species as it did for so many small marsupials. Introduced itself as a game animal and as a rabbit-control, it experienced a classic population explosion and found its easiest pickings among the native rather than the introduced fauna. At the same time the open grasslands were being altered by the grazing and trampling of sheep and cattle. In 1850 the Eastern Hare-Wallaby was a common animal in northwestern Victoria, New South Wales and eastern South Australia. But the last specimen ever recorded was taken in New South Wales, 48km (30 miles) north of Boligal in 1890.

GILBERT'S POTOROO

Macropodidae
Potorous gilberti
EXTINCT *c.* 1900
Western Australia

Gilbert's Potoroo
Potorous gilberti

An even more diminutive group of *Macropodidae* (Kangaroos) reminded the European settlers of rats, and are still widely known as Rat-Kangaroos, although Potoroo is the preferred name. Indeed, at an average length of 40cm (16in.) excluding the 16cm (6in.) tail, the Potoroos' resemblance to rats is very superficial. As with most of the smaller wallabies, the prospects for the remaining Potoroos are bleak (though they have been bred in zoos), and at least two, and probably three, species have become extinct.

Gilbert's Potoroo (*Potorous gilberti*), named for its discoverer the explorer-naturalist John Gilbert, is known to us only from the two specimens which he sent to the British Museum, and has never been recorded since. Characterized by its black tail and by the black stripe from nose to brow, this Potoroo was found only in the King George's Sound area of Western Australia, and the only information we have of its habits are contained in Gilbert's field notes, as published by John Gould: 'they are always found . . . amidst the dense thickets and rank vegetation bordering swamps and running streams. The natives capture them by breaking down a long passage in the thicket, in which a number of them remain stationed, while others, particularly old men and women, walk through the thicket, and by beating the bushes and making a yelling noise, drive the affrighted animals before them into the cleared space, where they are immediately speared . . .'

It was clearly not traditional hunting pressures, though, which affected Gilbert's Potoroo, but the encroachment on their habitat by man and livestock and the depredations of foxes and, perhaps, cats.

BROAD-FACED POTOROO

Macropodidae
Potorous platyops
EXTINCT *c.* 1908
Western Australia

John Gilbert also collected another Potoroo from the same area, though this one had a wider range. The Broad-faced Potoroo (*Potorous platyops*) was already scarce at that time and was known by the aboriginals as 'Mor-da', while in its other known habitat, the Margaret River area, the natives called it 'Warrack'. It was from this district that the last known specimen was taken in 1908.

This was a considerably larger animal than Gilbert's Potoroo (body 48.2cm, 19in.; tail 17.7cm, 7in.), with a distinctively wide face, 'beset with numerous yellowish-white hairs', and a brown tail. The natives of Margaret River described it as a slow-moving creature, easily caught, and blamed its disappearance on cats and on bush-fires started by the colonists.

ST FRANCIS ISLAND POTOROO

Macropodidae
Potorous sp
EXTINCT *c.* 1900
St Francis Island, Great Australian Bight

Cats were introduced deliberately in the late 1880s to keep down the swarms of Potoroos on St Francis Island in the Great Australian Bight, and they did their work so efficiently that by the turn of the century the 'Tungoos', as the colonists called them, were all gone. Not a single specimen or skeletal fragment was preserved for science and the St Francis Island Potoroo has never been fully described.

Professor Wood Jones visited the island and interviewed the single family who had established a small, subsistence-farm there and had

Broad-faced Potoroo
Potorous platyops

Eastern Hare-Wallaby *Lagorchestes leporides*

Eastern-Barred Bandicoot
Perameles fasciata

waged war on the Potoroos. His account contains almost all that is known about this species, and indicates vividly how vulnerable the Tungoos must have been when the cats came in: 'The animals do not seem to have formed burrows, but they lived in the undergrowth, and used frequently to hop into the homestead to take bread or other eatables thrown to them from the table. They do not appear to have been nocturnal; they do not seem to have been afraid of the human invaders of the island. Their only offence seems to have been that they had a liking for the garden produce of the family who settled on the island.'

EASTERN BARRED BANDICOOT

Peramelidae
Perameles fasciata
EXTINCT *c.* 1940
New South Wales and Victoria

The remaining Australian casualties belonged to a very different family of marsupials, the *Peramelidae* or Bandicoots, of which a large number of species are severely endangered by introduced predators and habitat destruction.

The mainly vegetarian Barred Bandicoots were non-burrowing forms, and hence most susceptible to fire and to the clearance and trampling down of cover. Both the Eastern (*Perameles fasciata*) and the Western (*Perameles myosura*) Barred Bandicoots are extinct. The former was found in New South Wales and Victoria, and was the more handsome of the two with its white underparts, yellow flanks, and head and back 'pencilled' with black and yellow. It had, besides, four pale vertical stripes on its rump.

WESTERN BARRED BANDICOOT

Peramelidae
Perameles myosura
EXTINCT *c.* 1910
Western Australia

The Western form (or 'Marl') was an altogether drabber animal, with a general mingling of black and pale brown above, underparts a 'dirty yellowish white', and one dark band across its sides, just in front of its thighs. It had a much wider range, extending from the southwest tip of Western Australia as far north as Shark Bay.

It should be noted that recent revisions of marsupial taxonomy have included both the Eastern and Western Barred Bandicoots in the full species *Perameles bouganville*, whose representatives survive on Bernier Island and one of the two Dorre Islands in Shark Bay.

PIG-FOOTED BANDICOOT

Peramelidae
Chaeropus ecaudatus
EXTINCT *c.* 1907
South Australia

A more distinctive relative was the Pig-footed Bandicoot (*Chaeropus ecaudatus*) which has only been reliably reported once since the turn of the century (in 1907) and which, according to Australian wildlife officials, is doomed to extinction even if a scattered few might still survive in the area south of Lake Eyre in South Australia. These grey-brown, graceful creatures, a bit smaller than a rabbit, had slender legs 'like miniature deer', long ears and a slightly crested tail. They are named from the resemblance of their forefeet to pigs' trotters, having only two properly developed toes. The hind feet consisted mainly of one greatly enlarged fourth toe. They were nocturnal, burrowing and strictly vegetarian. The Pitjanjarra natives claimed to have seen them around 1925, but there seems little doubt that they have indeed disappeared.

GREATER RABBIT-BANDICOOT

Peramelidae
Macrotis lagotis grandis
EXTINCT *c.* 1930
South Australia

The most attractive of all the bandicoots, and perhaps of all marsupials, are the Rabbit-Bandicoots or Bilbies. With their fly-away ears and their habit of sleeping sitting back on their tails, with their heads tucked between their forepaws and their ears folded forward along their long faces, the Rabbit-Bandicoots are entrancing, fay-looking creatures. The early Australian colonists apparently thought so too and often protected them. Indeed, Wood Jones commented on the unusual 'tolerance with which it was regarded by people whose hands may be justly said to have been against all animals'. He attributed this as much to the Rabbit-Bandicoots' undoubted usefulness in keeping down mice and insects, as to any aesthetic appeal they may have had. In any case, times changed, and with the growth of a lucrative fur market in Adelaide, affection was replaced by 'ruthless slaughter'.

A considerable number of Rabbit-Bandicoots also fell victim to a campaign against the proliferating, introduced rabbits. Many thousands of the small marsupials were killed or maimed in steel rabbit-traps. The fact that the Rabbit-Bandicoots were carnivorous may also have led some of them to take poisoned baits set out for dingoes, foxes and marsupial 'cats'.

Competition with rabbits for nesting-burrows, together with fox predation, may also have contributed to the complete disappearance of the largest known species Bilby, the Greater Rabbit-Bandicoot (*Macrotis lagotis grandis*: body 55cm, 22in.; tail 26cm, 10 in.; ears 7.7cm, 3in.). These animals seem to have become extinct sometime before 1930 and Wood Jones's remarks at that time probably tell most of the story: 'Not more than 30 years since it was usual for trappers, even in the immediate neighbourhood of Adelaide, to take more Bilbies of this type than rabbits in their traps.'

The Greater Rabbit-Bandicoot is sometimes known as the Nalpa Bilby, after its last stronghold near Lake Alexandrina in South Australia.

Western-Barred Bandicoot
Perameles myosura

Pig-footed Bandicoot
Chaeropus ecaudatus

Greater Rabbit-Bandicoot *Macrotis lagotis grandis*

Captain Maclear's Rat *Rattus macleari* Extinct *c.* 1900

Christmas Island Musk Shrew *Crocidura fuliginosa trichura* Extinct *c.* 1900

Rats and Bats

RODENTS, INSECTIVORES AND BATS

So many extinctions, of birds, mammals and reptiles, can be blamed on the rats which came off European ships, that it comes as something of a shock to find members of the rat family numbered among the casualties. The truth is that, in insular ecosystems especially, the rodent family was not markedly carnivorous and the rats and their relatives had established an harmonious niche with other life forms over the centuries. The Kiore, or Polynesian rat, was a fruit and seed-eater, much relished by the Polynesians as a food-source and introduced by them wherever they went. It colonized New Zealand, for instance, without apparently harming any other creatures, though it triggered a population explosion among the Laughing Owls who found in it a new and plentiful food supply. But when the European rats came the Kiore, and many other creatures, rapidly disappeared. The Carib and Arawak natives of the West Indies also transported (and probably bred selectively) edible rodents (Echimyidae) which were vegetarian and disappeared almost as soon as European rats (and the other villains, mongooses) were introduced.

Nowhere is the inoffensiveness and vulnerability of most rats

CAPTAIN MACLEAR'S RAT

Muridae
Rattus macleari
EXTINCT *c.* 1900
Christmas Island, Indian Ocean

Bulldog Rat *Rattus nativitatus* Extinct *c.* 1900

better illustrated than on Christmas Island, 200 miles south of Java (and not to be confused with the atomic wasteland in the Pacific). Thirteen years after a settlement had been established there to mine the rich phosphate deposits, two old world rats and one shrew had disappeared and the Black Rat (*Rattus rattus*) had inherited a paradise.

Captain Maclear's Rat (*Rattus macleari*), a long-tailed rodent about the size of our Brown Rat (48cm, 19in. with tail), was remarkable for the prominent, erect black hairs on its back. It was, otherwise, a 'grizzled rufous' colour with a paler belly. The first specimens were brought to England by the man for whom the rat is named, master of the survey-ship *Flying-fish*. This was in 1886 and the uninhabited island was described as a 104sq km (40sq mile) coral structure with a volcanic core rising to 366m (1200ft) and 'covered with jungle and forests'.

Ten years later the mining settlement was established at Flying-Fish Cove, the sole anchorage, and the following year Dr Charles Andrews' description of this, the island's most common mammal, is virtually our sole source of information but is remarkably evocative: 'In every part I visited it occurred in swarms. During the day nothing is to be seen of it, but soon after sunset numbers may be seen running about in all directions and the whole forest is filled with its peculiar querulous squeaking and the noise of frequent fights. These animals, like most of those found on the island, are almost completely devoid of fear, and in the bush if a lantern be held out they will approach to examine the new phenomenon.' He describes the nuisance they caused to the human settlers and continues: 'Their natural food appears to be mainly fruits and young shoots and to obtain the former they ascend trees to a great height . . . and frequently come into conflict with the fruit bats on the tops of the papaia-trees . . . In the daytime these rats live in holes among the roots of trees, in decaying logs, and shallow burrows. They seem to breed all the year round.'

Dr Andrews returned to Christmas Island in 1909 and found it little changed except for a few roads. But the Maclear's Rats were gone and so was the island's other rat.

BULLDOG RAT

Muridae
Rattus nativitatis
EXTINCT *c.* 1900
Christmas Island, Indian Ocean

This was the Bulldog or Burrowing Rat (*Rattus nativitatis*), a very different creature with a distinct ecological niche. It was much heavier than the climbing rat, with an overall length of 42cm (16.5in.) and a proportionately short tail. It was dark brown and thick furred and had 'a peculiarly small and delicate head'. Its strong, broad claws were adapted for digging. It was far more localized than Maclear's Rat, being restricted to the hilly interior of the island where it was first collected in 1887. Dr Andrew's account is, again, our principal reference:

'Though very numerous in places . . . it is very much less common than *macleari* . . . They seem to live in small colonies in burrows, often among the roots of a tree, and occasionally several may be found in the long, hollow trunk of a fallen and half-decayed sago-palm. The food consists of wild fruits, young shoots, and, I believe, the bark of young trees . . . It is a much more sluggish animal than *macleari*, and unlike it, never climbs trees . . . Strictly nocturnal, when exposed to daylight it seems to be in a half dazed condition. The Ross family in Christmas Island have given this species the name "Bull-dog Rat" and this has been adopted by the Malays.'

Andrews commented on the abundance of food and the absence of predators – 'the hawk and owl, which are the only possible enemies, feed mainly on birds and insects' – and noted that most specimens had 'a layer of fat from ½in. to ¾in. [1.2–1.9cm] thick over most of the dorsal surface of the body'. Nevertheless the Bulldog Rats, like Captain Maclear's Rat had all vanished by 1809, far too soon, one would think, for the immigrant Black Rats to have killed them off.

Andrews had his own theory about this and, considering the lack of immunity in most insular animals, he was probably right: 'The complete disappearance of two such common animals seems to have taken place within the last five or six years and to have been the result of some epidemic disease, possibly caused by some *trypanosome* introduced by the ship rats . . . a conclusion supported by an observation made by the medical officer, Dr MacDougal, who told me that some five or six years ago he frequently saw individuals of the native species of rats crawling about the paths in the daytime, apparently in a dying condition.'

CHRISTMAS ISLAND MUSK SHREW

Soricidae
Crocidura fuliginosa trichura
EXTINCT *c.* 1900
Christmas Island, Indian Ocean

Whether disease also killed off the Musk Shrew (*Crocidura fuliginosa trichura*), the only other indigenous quadruped on Christmas Island, is debatable. Andrews thought that the introduced cats were more likely responsible for the disappearance of the Christmas Island Musk Shrew but other authorities dispute this, on the grounds that cats will not eat shrews. However, as any rural cat-owner will testify, they have no inhibitions about killing them.

The dark grey insectivore (13.5cm, 5in.) was less unusual than the rats – being closely related to animals in Malaysia – and probably a more recent arrival. Nevertheless the Musk Shrew had been resident long enough to develop a distinctively long-haired tail and had certainly found itself in a favourable environment. Dr Andrews observed that it was common everywhere on the island and that 'at night its shrill squeak, like the cry of a bat can be heard on all sides. It lives in holes in rocks and roots of trees, and seems to feed mainly on small beetles'. On his second visit he found that, like Captain Maclear's Rat and the Bulldog Rat, it was apparently extinct.

South Australian Spiny-haired Rat
Rattus culmorum austrinus

SOUTH AUSTRALIAN SPINY-HAIRED RAT

Muridae
Rattus culmorum austrinus
EXTINCT *c.* 1850
Kangaroo Island

The South Australian Spiny-haired Rat (*Rattus culmorum austrinus*) was an old world rodent which seems to have become extinct in the mid-nineteenth century. It was possibly the victim of introduced European rats and foxes, but a more probable cause for its extinction was habitat destruction. The South Australian Spiny-haired Rat was first collected in 1841 on Kangaroo Island and was never reported after 1850. There are only seven known specimens of it in the world. About 27cm (10½in.) long, it was distinguished from other surviving subspecies by its longer hair and grey, rather than fawn colouring. Its relatives in Queensland and New South Wales are subject to massive population fluctuations and though there is some evidence that the South Australian form was once common, it may have been affected by human intrusion or introduced predators when it was at a low ebb and particularly vulnerable.

QUEMIS

Heptaxodontidae
Elasmodontomys obliquus
EXTINCT *c.* 1500
Puerto Rico
Quemisia gravis
EXTINCT *c.* 1700
Haiti

Across the world, on the islands of the Caribbean, an extraordinary profusion and variety of rodents had evolved, ranging from a creature the size of a black bear, (*Ambyloliza inundata*) – which may have died out before mankind arrived – to a 20cm, 8in. hamster-like Rice Rat which was eliminated in this century by the mongoose. Almost all of these Antillean rodents are extinct; most were victims of rats, mongooses, man and forest destruction; and in many cases our only knowledge of them comes from the investigations of cave deposits by H. E. Anthony, Gerrit Miller Jr and G. M. Allen. These descriptions are essentially summaries of those gentlemen's various publications.

The largest rodents from this area in historical times inhabited Haiti and Puerto Rico and have been identified by Gerrit Miller Jr. with the 'Quemi' described by Oviedo in his sixteenth-century account of Hispaniola. The Puerto Rican form (*Elasmodontomys obliquus*) died out quite early, but the Hispaniolan Quemi (*Quemisia gravis*) seems to have disappeared well after the Spanish settlement. Both had been a staple of the aboriginals' diet for centuries. From Oviedo's account, the Quemis were brown and about the size of the Paca, a burrowing rodent which survives today in South America.

HUTIAS

Capromyidae
CUBAN SHORT TAILED HUTIA
Geocapromys colombianus
EXTINCT *c.* 1850

CROOKED ISLAND HUTIA
G. ingrahami irrectus
EXTINCT *c.* 1600

GREAT ABACO HUTIA
G. ingrahami abaconis
EXTINCT *c.* 1600

HISPANIOLAN HEXOLOBODON
Hexolobodon phenax
EXTINCT *c.* 1600

HISPANIOLAN HUTIA
Plagiodontia spelaeum
EXTINCT *c.* 1750

Only slightly smaller were the hutias, a widespread family of rabbit-sized rodents, members of which survive in Cuba, Jamaica, the Bahaman Keys and on the mainland. The Cuban Short-Tailed Hutia (*Geocapromys columbianus*) seems to have died out within the last 150 years but before the coming of the mongoose. The other Cuban Hutias were tree-climbers and may have been less easy game for the European pot-hunters.

The surviving Bahaman Hutia (*Geocapromys ingrahami*) had two very close relatives on Crooked Island (*G. i. irrectus*) and Great Abaco (*G. i. abaconis*) which were mentioned by early European voyagers but whose extinction was probably more due to land clearance and brush fires than to direct hunting.

Hispaniola was home to a number of Hutia forms. Perhaps the earliest historical extinction among them was of the Hexolobodon (*H. phenax*) which seems to have been exterminated soon after European occupation. It was very similar to the Cuban and Bahaman *Geocapromys* genus just mentioned, but with some dental distinctions. This Hutia was described by Dr Miller from caves near St Michael in Haiti, as was the Hispaniolan Hutia (*Plagiodontia spelaeum*) an animal which apparently survived somewhat longer.

With this genus we are on far less hypothetical ground, for a larger form (*Plagiodontia aedium*) was described and illustrated by Cuvier in 1836. A hundred years later a living relative of intermediate size was discovered by W. L. Abbot in northeast Hispaniola and given the name *P. hylaeum* or Dominican Hutia. Then in 1947 Cuvier's Hutia was itself rediscovered in the island's south west. We can thus reconstruct the Hispaniolan Hutia with some confidence by reference to its surviving relatives.

It was a squat animal, about 18cm (7in.) long, with a naked 7.5cm (3in.) tail and fine, silky hair which was presumably grey, shading to buff with intermingled long black hairs. It is likely that it had paler underparts, as Cuvier's Hutia did, but this is guesswork: the Dominican Hutia is uniformly coloured. We can be certain that, like

Hispaniolan Hutia *Plagiodontia spelaeum* Extinct *c.* 1850

all this family, it had long and abundant moustachial hairs.

All these Hutias are slow (single pregnancy) breeders and poor climbers, and hence are vulnerable on two fronts. They were known as 'house rats' ('Rat Cayes') because, though nocturnal, they favoured settled areas. They ate fruit and seeds and their own flesh was so esteemed that even in Cuvier's time the family had been hunted almost to extinction. The surviving species are drastically threatened by mongooses.

ISOLOBODONS

Capromyidae
PUERTO RICAN ISOLOBODON
Isolobodon portoricensis
EXTINCT *c.* 1700
HAITIAN (HISPANIOLAN) ISOLOBODON
Isolobodon levir
EXTINCT *c.* 1550
HISPANIOLAN NARROW-TOOTHED HUTIA
Aphaetraeus montanus
EXTINCT *c.* 1600

Of much the same size and habits, and equally popular as a food, were the Puerto Rican and Hispaniolan Isolobodons. The Puerto Rican animal (*Isolobodon portoricensis*) was the larger and more common and its remains are so widespread, and unvarying, that we must assume that the aboriginals transported it and very probably bred it (perhaps selectively, for size) in semi-captivity. It is presumed to have originated in Puerto Rico because its remains are most numerous there.

The Haitian Isolobodon (*I. levir*) had been a favoured prey of the extinct giant Barn Owl (*Tyto ostalaga*) but became scarce several centuries ago. At this stage the Indians of Hispaniola and the Virgin Islands seem to have turned to the larger, imported isolobodon (along with the domesticated guinea pig), as a primary food source.

A more highly developed Hispaniolan octodont rodent, the Narrow-Toothed Hutia (*Aphaetraeus montanus*), was contemporary with, and closely related to the isolobodons, but was never mentioned by the Spaniards nor – so far as kitchen middens indicate – exploited by the aboriginals. There is no evidence for its survival past the initial European settlements and the spread of rats from the ships, but it is not impossible that it was alive much more recently than that as, apparently, were two agouti-like contemporaries known now only from cave-deposits on Puerto Rico.

PUERTO RICAN 'AGOUTIS'

Echimyidae (Heteropsomynae)
Heteropsomys insulans
Homopsomys antillensis
EXTINCT *c.* 1750
Puerto Rico

Perhaps 25cm (10in.) long, and hence slightly smaller than the surviving agoutis of Trinidad (which are themselves gravely threatened by mongooses and the destruction of watersheds), these Puerto Rican 'Agoutis' (*Heteropsomys insulans* and *Homopsomys antillensis*) were very closely related to each other, but Dr Anthony believes they were somewhere between agoutis and spiny rats and has proposed a new subfamily, Heteropsomynae for them – although other authorities do not agree. Both of these rodents seem to have become extinct by 1750.

PUERTO RICAN CAVIOMORPH

Heptaxodontidae
Heptaxodon bidens
EXTINCT *c.* 1600
Puerto Rico

Puerto Rico was the most remarkable of all the Caribbean islands for the variety of rodent genera which overlapped in time and habitat. The Puerto Rican Caviomorph, *Heptaxodon bidens*, was one such rodent. It was about the size of a ground-hog (30.5cm, 12in.) but is known only from jaw fragments datable from about the time of European settlement. Despite the remarkably detailed inferences that experts can draw from skeletal fragments, all we can really point to is the former wide range of the *Heptaxodon* in those forested areas of the Puerto Rican plateau which have now been completely replaced by plantations. Whatever the Heptaxodon may actually have looked like, the principal cause of its extinction is very clear.

The West Indian Spiny Rats had much the same distribution as the hutias. According to Gerrit Miller Jr, four species of spiny rats have become extinct – two each from Hispaniola and Cuba. The Hispaniolan Spiny Rats (*Brotomys voratus* and *Brotomys contractus*) were described from the discovery of bones in an aboriginal midden near Santo Domingo, and Dr Miller wrote: 'I have little doubt that this animal was the Mohuy described by Oviedo as . . . "somewhat smaller than the hutia, its colour is paler and likewise grey". This was the food esteemed by the caciques and chiefs of this island, and the character of the animal was much like the hutia except that the hair was denser and coarser . . . and very erect.'

It seems that these two spiny rats (distinguishable mainly by the peculiarly narrow palate of *B. contractus*) did not long survive the coming of Europeans and their rats, unlike the Cuban forms (*Boromys offella* and *Boromys torrei*). The Cuban Spiny Rats were distinguishable from each other only by size and although both were quite common in the early nineteenth century, they seem to have become extinct about 1870. What exact combination of predation and interference caused their extinction cannot now be determined.

SPINY RATS

Echimyidae
Brotomys voratus
Brotomys contractus
EXTINCT *c.* 1600
Hispaniola
Boromys offella
Boromys torrei
EXTINCT *c.* 1870
Cuba

More like rats as we understand them are the cricetine Rice Rats of Jamaica (*Oryzomys antillarum*) and St Vincent which, unlike most Caribbean rodents, were not hystricomorphs. The Jamaican Rice Rat was still alive in 1877, but by then the mongoose had arrived. The 25cm (10in.) long creature (half of this was tail) was described by P. H. Gosse in 1845 as a 'field mouse' of 'a beautiful reddish colour with a milk-white belly', which lived 'chiefly about the hollow roots of large trees' and by river banks. It was already scarce in his time.

JAMAICAN RICE RAT

Cricetidae
Oryzomys antillarum
EXTINCT *c.* 1880
Jamaica

Jamaican Rice Rat *Oryzomys antillarum*

ST VINCENT RICE RAT

Cricetidae
Oryzomys victus
EXTINCT *c.* 1900
St Vincent, West Indies

The only known specimen of the St Vincent Rice Rat (*Oryzomys victus*) was collected in 1897. It was a smaller animal (20.5cm, 8in.) than the Jamaican species and was known to be a forest dweller. A contemporary observer blamed this rodent's disappearance squarely upon the 'indiscriminately destructive' mongoose.

'MUSK RATS'

Cricetidae
MARTINIQUE MUSK RAT
Megalomys desmarestii
EXTINCT *c.* 1900
ST LUCIA MUSK RAT
Megalomys luciae
EXTINCT *c.* 1880
BARBUDA MUSK RAT
Megalomys audreyae
EXTINCT *c.* 1600
West Indies

The islands of Martinique, St Lucia and Barbuda harboured a related genus of rodents, *Megalomys*: black and white animals known as 'Musk Rats'. They were much larger animals, measuring up to 71cm (28in.) overall, and Du Tertre, in his *Histoire Generale* . . . of 1654 mentions finding one in a snake's belly on Martinique which was 'almost as big as a cat'. The natives singed their fur off and boiled them twice to remove the strong musky odour. The Barbuda Musk Rat (*Megalomys audreyae*) probably became extinct as early as 1600, while the St Lucia species seems to have survived until about 1880. The Martinique form survived the longest, disappearing after the island's volcanic eruption of 1902 with its clouds of poisonous gas.

CARIBBEAN INSECTIVORES

Nesophontidae
PUERTO RICAN NESOPHONT
Nesophontes edithae
CUBAN NESOPHONT
Nesophontes micrus
CUBAN LONG-NOSED NESOPHONT
Nesophontes longirostris
EXTINCT *c.* 1650
HISPANIOLAN NESOPHONT
Nesophontes paramicrus
LESSER HISPANIOLAN NESOPHONT
Nesophontes hypomicrus
LEAST HISPANIOLAN NESOPHONT
Nesophontes zamicrus
EXTINCT *c.* 1930
West Indies

Puerto Rico, Cuba and Hispaniola nurtured, as well as the amazing group of rodents already discussed, a primitive genus of insectivores named *Nesophontidae* by Dr Anthony, who first discovered remains in the Cathedral Cave near Morovis, Puerto Rico. On average they were about the size of a chipmunk (25cm, 10in., overall), with long, flexible snouts, narrow skulls, small eyes and short legs. The males seem to have been considerably larger than the females, an unusual phenomenon in insectivores, as is the separation of the pubic bones and of the tibiae and fibulae. G. M. Allen writes 'No living insectivore has such an assemblage of generalised characters' and it has been proposed that they were ancestral to the moles. The Puerto Rican form (*N. edithae*) was the largest and, like the Cuban forms (*N. micrus* and *N. longirostris*), disappeared exactly at the time that European rats arrived. The three Hispaniolan Nesophonts (*N. paramicrus*, *N. hypomicrus* and *N. zamicrus*) were alive very much more recently. Gerrit Miller Jr expressed hope in 1930 that they might still survive, but his hopes have not been justified by any recent reports, and their habitat has meanwhile been much more disturbed.

From the same three islands comes evidence of yet another group of primitive quadrupeds whose presence, combined with the genera already described, must have made these Caribbean enclaves almost intoxicatingly opulent to the first aboriginal hunter-gatherers to beach their canoes there. These creatures were of the order Edentata, or Ground Sloths, diminutive cousins of the famous *Megatherium*, the Giant Patagonian Ground Sloth, with a size range of from 30 to 500lb. Like the South American monster, they are known to have been hunted extensively by the native peoples.

It must be emphasized that these animals became extinct long before the period covered by this book, but no picture of the Caribbean would be complete without mention of these vanished forms: not just to demonstrate the amazing diversity of mammals which flourished and developed there, but also as a reminder that modern, western Man was not the first of his kind to invade new territories and exploit the edible fauna to the point of extinction.

Widespread habitat destruction in the Caribbean has been responsible for the loss of a number of species of bats. Though bats are relatively immune to introduced predators they are vulnerable in two particular areas. One is in their roosting, for their colonies depend on special conditions of temperature and humidity, whether in caves or hollow trees, and the number of suitable sites is always limited and finite. The other is in their food-supply. Over the centuries, some species have adapted to a narrow and specialized diet which can easily be disturbed, diminished or destroyed. The Caribbean casualties were all fruit-eaters, and their disappearance was almost certainly caused, for the most part, by the clearance of native fruiting trees. Theoretically it only needs the destruction of one fruit species to set off a devastating chain-reaction, because trees fruit seasonally and if one month alone becomes barren a dependent species can starve, even if there is abundant food during the rest of the year.

JAMAICAN LONG-TONGUED BAT

Phyllostomatidae
Reithronycteris aphylla
EXTINCT *c.* 1900
Jamaica

Five of the six extinct bats were members of the family Phyllostomatidae, leaf-nosed bats of the New World which seem to have developed in South America and are extremely widespread, constituting 52 genera. The Jamaican Long-tongued Bat (*Reithronycteris aphylla*) is known only from one specimen, a male taken sometime before 1898 and preserved in alcohol at the Institute of Jamaica. Animals preserved in this way can present a misleading appearance and we can only say that *at present* it is a light yellowish brown and measures nearly 9cm (3½in.) overall. It is unusual for its family in that its nose-leaf is rudimentary and 'appears as a short, pig-like snout, set off by an encircling groove'. It is short-eared and has a strikingly big hind foot with strong claws; while it is remarkable

Martinique Musk Rat *Megalomys desmarestii*

internally for its almost sub-divided brain. But since it is the only known specimen of its genus, there are no living forms by which one might guess at the significance of all these features. Absolutely nothing is known about its original habitat or life-style, or even which fruits it depended upon.

PUERTO RICAN LONG-TONGUED BAT

Phyllostomatidae
Phyllonycteris major
EXTINCT *c.* 1850
Puerto Rico

We know more about the subfossil, related Long-tongued Bats of Puerto Rico (*Phyllonycteris major*) and Haiti, because a living form (*P. poeyi*) survives in some numbers in Cuba and can be used as a point of reference. These were small bats, narrow-headed and almost totally dependent on fruit pulp and juice, though their long, protrusible tongues were capable of 'milking nectar and pollen too'. They had extremely weak teeth and the nose-leaf was small.

The Puerto Rican Long-tongued Bat is known from only one cave investigated by Dr Anthony but its remains were numerous (over 60 skulls) and the supposition is that, like its surviving Cuban relative, it was a cave-dweller, and a very specialized one at that, requiring 'wet, ill-ventilated caves, stiflingly hot in the summer (especially just after the young are born)'. In Cuba the young are described as 'pink, almost hairless . . . of different sizes, hanging to the roof and scattered over much of its surface'. This description is by Dr Miller of the Smithsonian Expedition, who also commented on the 'astounding noise of the bats' wings in the cave'.

HAITIAN LONG-TONGUED BAT

Phyllostomatidae
Phyllonycteris obtusa
EXTINCT *c.* 1900
Hispaniola

The Haitian form (*Phyllonycteris obtusa*) was markedly smaller (7cm, 2½in., against 9cm, 3½in.) and its teeth showed some variations, but we can assume that its life-style was almost identical. It may well have had a wider distribution, however, since Dr Miller found its remains in three quite distinct cave districts. In one cavern (at Diquini) a skull was retrieved from owl pellets whose date was so recent that Dr Miller was for a while quite hopeful that this bat might still survive. Certainly its known predators (snakes and owls) have not increased in number, nor has there been any climatic shift in that region or evidence of disease among bats. There is no natural explanation for the Haitian Long-tongued Bat's extinction.

However, a clue to its disappearance may lie in the fact that most bones were found in the 'Crooked Cave' near the Atalaye plantation. It is easy to imagine that land clearances there may have interrupted the seasonal fruit sequence, as outlined earlier.

LESSER FALCATE-WINGED BAT

Phyllostomatidae
Phyllops vetus
EXTINCT *c.* 1750
Cuba

Dr Anthony's investigations of a Cuban cave turned up evidence in owl-pellets of another small fruit-eating bat, the Lesser Falcate-winged Bat (*Phyllops vetus*) which has apparently been extinct for at least 200 years. Curiously, a very similar bat (*P. falcatus*) survives on the island to this day and though the only difference is size (the living form is considerably larger) this must have been a significant factor in the extinction. Possibly the smaller bat was more susceptible to owl predation and its extinction may not be attributable to anything but natural selection. It was presumably, like the living form, a tree-dweller and just possibly favoured a roosting tree which has been destroyed. Climatic change may have played a part too: a bat's survival in this drier eastern part of Cuba is described as 'precarious' anyway.

The indefatigable Dr Anthony has also named the Puerto Rican Long-nosed Bat (*Monophyllus plethodon frater*) from the Morovis Cave. Here is another instance of two close relatives living side by side and only one (*M. redmani portoricensi*) surviving, but in this case the smaller form proved the hardier. Characterized by an unusually projecting tail, the Long-nosed Bat was presumably more affected by forest clearances and fires than its cousin. The Puerto Rican Long-nosed Bat had a characteristic long snout, surmounted by a small, lancet-shaped nose-leaf and a long, extensible tongue for feeding on fruit juices and also, perhaps, on nectar or pollen. It could be distinguished from other forms by its longer, more protruding tail. It probably became extinct by the mid-nineteenth century for a number of collectors who have searched its relatively restricted habitat since that time have seen no sign of it.

PUERTO RICAN LONG-NOSED BAT

Phyllostomatidae
Monophyllus plethodon frater
EXTINCT *c.* 1850
Puerto Rico

Finally, from the family of Long-legged Bats (Natalidae), there was the Cuban Yellow Bat (*Natalus primus*), also named by Anthony from cave deposits of fairly recent date in the Cueva de los Indios near Daiquiri, Cuba. It was named, by guesswork, from the colouration of its surviving relatives on the Bahamas, Curaçao and the mainland of South America. The Cuban Yellow Bat was the largest and most heavily built of its family, which are distinguished by their long, slender legs and funnel-shaped ears. But the extinct form has been described from teeth only and further comment would be merely guesswork. The Cuban Yellow Bat had probably become extinct by 1850 for it has not been seen by any naturalists or collectors working in Cuba since that date. With such meagre remains the date of extinction must be conjectural.

CUBAN YELLOW BAT

Natalidae
Natalus primus
EXTINCT *c.* 1850
Cuba

Jamaican Long-tongued Bat *Reithronycteris aphylla*

Part Three

REPTILES, AMPHIBIANS AND FISH

Rodriguez Greater (Saddleback) Tortoise *Geochelone (Cylindraspis) vosmaeri* Extinct *c.* 1800

Reptiles and Amphibians

TORTOISES, SNAKES, LIZARDS AND FROGS

RODRIGUEZ GIANT TORTOISES
Testudinidae
GREATER (SADDLEBACK) TORTOISE
Geochelone (Cylindraspis) vosmaeri
LESSER TORTOISE
Geochelone (Cylindraspis) peltastes
EXTINCT *c.* 1800
Rodriguez and satellites

The spectacular giants of the Age of Reptiles stand in popular mythology for obsolescence: the first victims of the ruthless winnowings of evolution. Their descendants are objects of mixed repulsion and fascination for us. They seem creatures of a half-life, sub-animal and with some of their ancestors' fearful power. Crocodiles, snakes and two lizards apart, though, the tortoises are the only giants whose time has overlapped with ours and they are, perhaps, the saddest victims of exploitation ever recorded.

In historical times the giant forms of the genus *Geochelone* were confined to the Galapagos Islands off Ecuador, and to the Mascarenes and Seychelles in the Indian Ocean. All the Indian Ocean giants were gone by the early-nineteenth century except the Aldabran Tortoise; while four of the 14 Galapagoan forms are extinct and five survive marginally, wholly or partly in captivity. Their downfall was that they made delicious eating and were a constant source, on the longest sea voyages, of fresh meat.

All three Mascarene islands and most of their satellites were densely populated by tortoises when discovered, and the Rodriguez

Rodriguez Lesser Tortoise *Geochelone (Cylindraspis) peltastes* Extinct *c.* 1800

species, especially, were in prodigious abundance. Yet our knowledge of these animals is based very largely on fossil evidence: we are not even positive how many forms there were. François Leguat, the Huguenot castaway, speaks of three varieties, but modern authorities are only certain about two: the true giant (*Geochelone vosmaeri*, 85cm, 33in.) and the smaller form (*G. peltastes*, 42cm, 17in.). They are distinguished more by adaptive variations than size, however. The larger tortoise was a 'saddleback', having an upturned front to its carapace (upper shell) which enabled it to raise its head more and browse off higher herbage than its relative could. The smaller animal's carapace was domed and sloped directly to the ground, restricting it to low-grazing. The two, originally very similar, had thus adapted to share out the available island vegetation.

Leguat's journal recorded: 'There are such plenty of Land-Turtles on this Isle, that sometimes you see 2–3000 of them in a Flock; so that you can go above a hundred paces on their backs without setting foot on the ground. They meet together in the evening in shady places, and lie so close, that one would think those spots were paved with them.' The largest, he said, weighed about 45kg (100lb) and he rhapsodized about their meat, especially the liver which was so delicious that 'you could say it always carries its own sauce with it'.

Though pirates and occasional Dutch naval ships had been taking tortoises from Rodriguez for some time, it was not until Leguat's memoirs were published in 1708 that the island came to be regarded as a meat reservoir for the French and English navies. Thereafter, ships called regularly to take all the tortoises they could hold.

By this time the tortoises on the neighbouring colony of Réunion were becoming scarce and the herds on Mauritius had probably vanished. There was now some rivalry about rights to the tortoises, and while the short-lived colony on Rodriguez itself survived the settlers were strictly held to a quota of the reptiles until their introduced cattle had multiplied. When the colony failed in 1730, the French East India Company decreed that Mauritians could only take tortoises from the small northern satellites, Round and Flat Islands, and that even ships on the India route were to be rationed: 'the captains must be forbidden to send out their boats to collect tortoises without informing the island commandants and stating the numbers they require'. The company was desperate to conserve the meat-supply until a port of call could be established in the Atlantic.

All these measures were ignored. When Mahé de Labourdonnais arrived as Governor General of Mauritius and Réunion he found his company's own ships were the worst plunderers of Rodriguez. Labourdonnais recolonized Rodriguez with a small group of soldiers, lascars and slaves, the plan being to gather the tortoises, ship them and 'stable' them on Mauritius. They sent 10,000 tortoises off annually. Some shiploads were of 6000 tortoises and on several occasions three-quarters of the cargo perished. The last big haul was in January 1768 when *L'Heureux* took off 1215 'carosses' (the largest size).

In 1791 the last overseer, Jean de Valgny, died on Rodriguez, a virtual castaway dependent for food for himself and two slaves largely on the generosity of visiting ships. For the tortoises were gone and, with them, France's interest in Rodriguez. The last tortoises ever seen there were two down a ravine in 1795.

The tortoises on Mauritius itself had suffered much earlier, not only from human predation but from introduced pigs and swarms of rats. Their early disappearance means that little documentation has survived. The first mention is from 1630 when the Englishman Sir Thomas Herbert wrote of 'land Tortoyses (so great that they will creepe with two men's burthen, and serve more for sport than service or solemne Banquet) . . .' and showed a prophetic awareness of their significance: 'The Ile has no humane inhabitants. Those creatures that possesse it, have it on condition to pay tribute (without exception) to such ships as famine, or foule weather force to anchor there.'

Herbert reflects the timeless British distrust of exotic dishes, and in contrast to the French encomia on tortoise-flesh he calls it 'odious food' and 'better meat for Hogs than Men'. The French settlers thought otherwise and fifty years later were slaughtering and salting down hundreds of the tortoises (whose survival powers aboard ship were not yet appreciated) or rendering them down. It took four to five animals to yield 45kg (100lb) of fat.

In 1673 Hubert Hugo, a French official, had to go to offshore islands to find tortoises in any quantity and his reports mention the carnage caused by pigs among the young tortoises, as well as their rooting up of eggs from the sand. The Mauritian giants were gone by 1760 and the herds that live there now are mostly descendants of introductions from Aldabra.

As with Rodriguez, there were at least two forms of tortoise on Mauritius which may have utilized different vegetations. The Domed Mauritian Giant (*G. inepta*) was probably a ground feeder not unlike the Rodriguez Lesser Tortoise, while the longer-established High-fronted Tortoise (*G. trisserata*) had a flatter shell, a longer neck, and a unique palatal feature which suggests its diet: it had three instead of the usual two biting ridges on each side of its mouth, giving it potentially greater plucking and chewing power, if feeding off shrubs and the lower leaves of trees.

MAURITIAN GIANT TORTOISES

Testudinidae
DOMED TORTOISE
Geochelone (Cylindraspis) inepta
HIGH-FRONTED TORTOISE
Geochelone (Cylindraspis) trisserata
EXTINCT *c.* 1700
Mauritius and satellites

Domed Mauritian Giant Tortoise
Geochelone (Cylindraspis) inepta

High-fronted Mauritian Giant Tortoise
Geochelone (Cylindraspis) trisserata

The Réunion Giant Tortoises (*G. indica* and *G. borbonica*) were first described by Du Quesne in 1650, 138 years after the island's discovery. He wrote: 'There are vast numbers of them: their flesh is very delicate and the fat better than butter or the best oil, for all kinds of sauces . . . the biggest ones can carry a man with greater ease than a man can carry them.'

By 1688 the tortoises were regarded as the staple food of the Réunion colonists and, as on Mauritius, the introduced pigs were gorging on the hatchlings and eggs. As the history of the Rodriguez animals has detailed, the Réunion stocks were soon depleted. A few survived until about 1732, when an anonymous French chronicler pronounced them 'utterly destroyed'. However, some lived on in captivity until as late as 1773.

REUNION GIANT TORTOISES

Testudinidae
Geochelone (Cylindraspis) indica
EXTINCT *c.* 1760
Geochelone (Cylindraspis) borbonica
EXTINCT *c.* 1773
Réunion

The giant tortoises of Aldabra, the Seychelles and the Amirante Islands are incompletely described, and we do not know how many forms there may have been. The introduced herds which survive on Mauritius may contain hybrids of various forms, but are primarily imports of the surviving Aldabra Giant (*G. gigantea*).

The Aldabra form survived partly because the remote spot was not

MARION'S (SEYCHELLES) GIANT TORTOISE

Testudinidae
Geochelone (Cylindraspis) sumeirei
EXTINCT *c.* 1918
Seychelles and Amirante Islands

settled until 1889 and partly through government measures, tolerant settlers and the efforts of individuals like Walter Rothschild. The governor reported gloomily in 1911: 'In a wild state practically all the young are killed by cranes, rats and wild cats' but today the tortoises seem safe and some are exported every year.

Though there is some debate concerning the Seychelles tortoises, one type – *Geochelone sumeirei* – has been given specific status. This form is based upon 'Marion's Tortoise', which was taken from the Seychelles to Mauritius by Chevalier Marion de France in 1766 as a gift for the Port Louis garrison. This mascot animal survived until 1918, when it died as the result of a fall. By that time it had been in captivity for at least 152 years.

GALAPAGOS GIANT TORTOISES

BARRINGTON ISLAND TORTOISE
Geochelone (elephantopus) sp.
EXTINCT *c.* 1890
CHARLES ISLAND TORTOISE
Geochelone (elephantopus) elephantopus
EXTINCT *c.* 1876
ABINGDON ISLAND TORTOISE
Geochelone (elephantopus) abingdoni
EXTINCT *c.* 1957
NARBOROUGH ISLAND TORTOISE
Geochelone (elephantopus) phantastica
EXTINCT *c.* 1906
Galapagos Islands

The Galapagos Islands in the Pacific Ocean will always be associated with Charles Darwin's insights into evolutionary process. Indeed, the generative spark for his theories, and much supporting evidence, came from his study of the differing races or species of giant tortoise on each island. The Galapagos were, in fact, named for their tortoises, from the Spanish 'galapagar'. Fourteen subspecies of the Galapagos Tortoise (*Geochelone elephantopus*) can be distinguished with some confidence, differing from each other in shape, colour, thickness of carapace, size, and length of limbs and necks. Limited experiments in zoos and on the islands suggest that the different forms can interbreed but with a much diminished fertility rate.

First discovered by Europeans in 1535, the Galapagos had become, by the late-seventeenth century, a regular port of call for whalers, naval vessels and buccaneers. It is from the 1684 *New Voyage Round the World* by the pirate naturalist William Dampier that we get our first full description of the tortoises; 'so numerous that 500 or 600 men might subsist on them alone for several months without any other sort of provision. They are extraordinarily large and fat, and so sweet that no pullet eats more pleasantly. One of the largest of these creatures will weigh 150–200lb [68–90kg] and some of them are two foot or two foot six inches over the carapace or belly.'

The American sailors and whalers had more idea than the French of the tortoises' powers of survival and took huge numbers alive. They called the meat 'Galapagos mutton' and though often the animals were simply shoved in piles into a ship's hold, some captains (like Amaro Delano in 1800) kept them on deck and took prickly pear plants with them as fodder. More typical treatment is reflected in the 1823 memoirs of Captain Benjamin Morrel: 'I have had these animals on board my own vessels from 5–6 months without their once taking food or water . . . they have been known to live on some of our whale ships for 14 months without any apparent diminution of health or weight.'

A truly pathetic account of such cargoes appears in the 1812 *Journal of a Cruise Made to the Pacific Coast* by Admiral Porter of the US Navy, who was actually the first man to comment on the differences between the tortoises of the various islands. He had captured two British ships supplied with 'elephant tortoises' but they had thrown them overboard to clear the decks for action: 'A few days afterwards we were so fortunate to find ourselves surrounded by about 50 of them which were picked up and brought on board, as they had been lying in the same place where they had been thrown over,

Charles Island Tortoise
Geochelone (elephantopus) elephantopus

Abingdon Island Tortoise
Geochelone (elephantopus) abingdoni

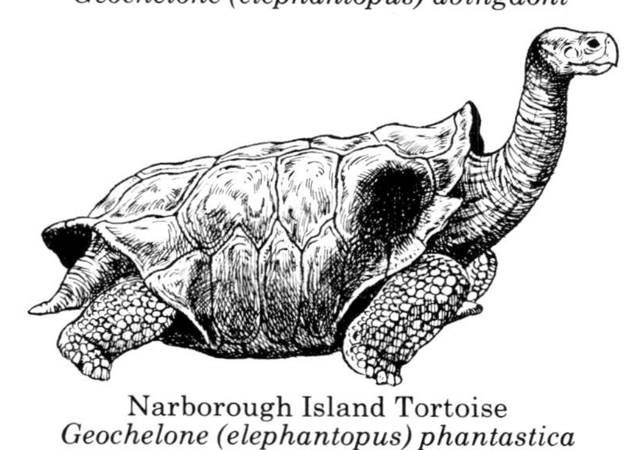

Narborough Island Tortoise
Geochelone (elephantopus) phantastica

incapable of any exertion in that element, except that of stretching out their long necks.' Porter had himself taken on 14 tons (13.8 tonnes) of the creatures, averaging 27kg (60lb) each, a few days before.

Porter distinguished the 'plump, black' James Island species from the saddlebacks of Hood, and Charles Darwin in his turn distinguished only two or three with any confidence and by his time the numbers were already in decline. The spread of human settlements, the ravages of dogs and pigs and the competition of goats which browsed off the cacti, berry-bushes and mosses, were even more important factors than the human meat-gatherers.

By 1890 the Barrington Island Tortoise (*Geochelone elephantopus sp.*) had been hunted out without being adequately described; while the Charles Island form (*G. e. elephantopus*) had disappeared even earlier, about 1876. This was somewhat like the 'plump, black' Hood Island tortoises but was flatter-backed, broader and even shinier, and measured up to 115cm (45in.).

The two pronouncedly saddlebacked forms from Abingdon (*G. e. abingdoni*) and the Narborough (*G. e. phantastica*) are probably gone too. The former, a thin-shelled, yellow-faced tortoise which reached 100cm (40in.) in length has not been seen alive since 1957 when goats were introduced though a solitary male of the species is kept by the Charles Darwin Research Station; while the Narborough Tortoise, with its distinctive, splayed shell-margins is only known from one specimen taken in 1906 (88cm, 34.5in.).

Ironically, some of these extinct forms, or hybrids of them, might exist in zoos, unidentified. Two well-intentioned American expeditions took off 436 tortoises early in this century and distributed them to zoos in the hope of founding breeding colonies. The hopes have not been justified, and the zoos' own records are often inaccurate. These animals can live for a hundred years and the possibility of 'restoring' some species might still exist if the time and money could be found to identify and collect living specimens from around the world.

One of the most successful of reptilian orders is the Serpentes, but even here we find Man's activities have resulted in five extinctions during the last few decades.

Four of these snakes were West Indian species of the Colubridae family, and were listed as extinct in the 1975 Schwartz and Thomas *Checklist of West Indian Amphibians and Reptiles*. In each case, the introduction of the mongoose was the prime factor, although direct human persecution had its effect – like nearly all West Indian snakes, they were killed on sight.

The Jamaican Tree Snake (*Alsophis ater*) before the introduction of the mongoose was evidently found throughout Jamaica. The related St Croix Tree Snake (*A. santicrucis*) was similarly common on the small island of St Croix in the American Virgin Islands. Both have become extinct within the last two or three decades.

The Martinique Racer (*Dromicus cursor*) and the St Lucia Racer (*D. ornatus*) were not restricted to those islands, but were found on at least one of their satellites. In both cases, the last specimens were collected on these islets: the Martinique Racer on Rocher de Diamont in 1962, and the St Lucia Racer on Maria Island in 1973.

The fifth and most recent snake extinction occurred on a tiny

WEST INDIAN SNAKES

Colubridae
JAMAICAN TREE SNAKE
Alsophis ater
EXTINCT *c.* 1960
ST CROIX TREE SNAKE
Alsophis sancticrucis
EXTINCT *c.* 1950
MARTINIQUE RACER
Dromicus cursor
EXTINCT *c.* 1962
ST LUCIA RACER
Dromicus ornatus
EXTINCT *c.* 1973
West Indies

ROUND ISLAND BOA

Boidae
Bolyeria multocarinata
EXTINCT *c.* 1980
Round Island, Mauritius

satellite formed of volcanic tuff (basalt) lying at about the 100 fathom line just offshore of Mauritius. This is Round Island, characterized in a recent article by Stanley Temple as having 'more species of threatened plants and animals per acre than any other piece of land in the world'. This 1.5sq km (374 acre) dot in the ocean supports six endemix or indigenous lizards, four trees in the same category, and two endemic snakes. One of these snakes, the Round Island Boa (*Bolyeria multocarinata*) is effectively extinct while the other, the Keel-Scaled Boa's (*Casarea dussumieri*) only hope of survival lies in seven specimens taken from the island in 1974 and housed by the Jersey Wildlife Preservation Trust.

'Effective extinction' means the existence of one, or even two or three specimens with no hope of reproducing. An intensive search in 1974 revealed only one Round Island Boa, a brownish snake about one metre (39in.) long, adapted with its pointed snout and cylindrical head and body for burrowing in loose tuff topsoil and palm litter. The loss of such habitat is explained by the release of rabbits and goats in 1844. By 1900 these aliens had so browsed off the herbage and true tree-seedlings that massive erosion set in, triggered by the cyclones and seasonal rains of those parts. Round Island is now furrowed by gullies up to 9m (30ft) deep. This led the browsers to the remaining palm-tree seedlings so that only a few adult palms are left, including only five Hurricane Palms and 16 Bottle Palms. A campaign to remove goats and rabbits is almost finished and, too late for the Boa, the island may, over the years be somewhat restored.

Réunion Skink
Gonglyomorphus bojerii borbonica

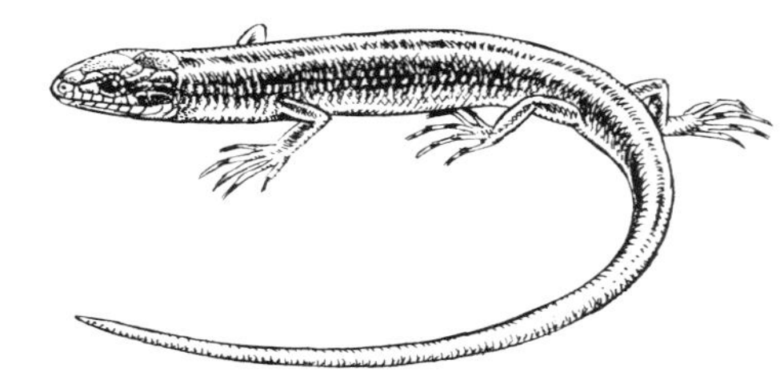

RÉUNION SKINK

Scincidae
Gonglyomorphus bojerii borbonica
EXTINCT *c.* 1880
Réunion

Among the reptiles to disappear from neighbouring Réunion was the once common skink (*Gonglyomorphus bojerii borbonica*) which frequented both shores and cultivated fields where it lived among grass and stones. Tame and inquisitive, even climbing over observers' legs, it walked briskly, jerking its head from side to side and flickering its tongue. It was bluish-green to brown in colour, marked from head to tail with three dark stripes, and measured about 15cm (6in.) of which more than half was tail. It was an egg-laying lizard, and this may have made it more vulnerable to predation by rats and mongooses; while attempts to keep it in captivity failed completely, even when it was offered its staple foods of insects and centipedes. The Réunion Skink has never been seen in this century.

RODRIGUEZ DAY GECKO

Gekkonidae
Phelsuma edwardnewtoni
EXTINCT *c.* 1920
Rodriguez and satellites

Though Rodriguez has a large number of satellite islands which have supported reptiles, they have not proved adequate refuges, since most of them, even when they lie as much as 2.5km ($1\frac{1}{2}$ miles) offshore, like Booby Island, have long been infested with rats. Two geckos have disappeared from Rodriguez and its satellites.

The Rodriguez Day Gecko (*Phelsuma edwardnewtoni*) was a brightly coloured lizard, 22cm (9in.) long – of which just over half was tail – and bright green above, stippled with vivid blue dots and with a striking yellow neck. Most of the early colonists and visitors commented on it and its habit of clinging to the branches of *Latania* palms, three or four feet above the ground. François Leguat emphasized that it was active by day, but had a number of natural enemies: if one was ever dislodged from a tree it was instantly pounced upon by one of the (extinct but undescribed) flightless herons which stood

Round Island Boa *Bolyeria multocarinata* Extinct *c.* 1980

Rodriguez Day Gecko
Phelsuma edwardnewtoni

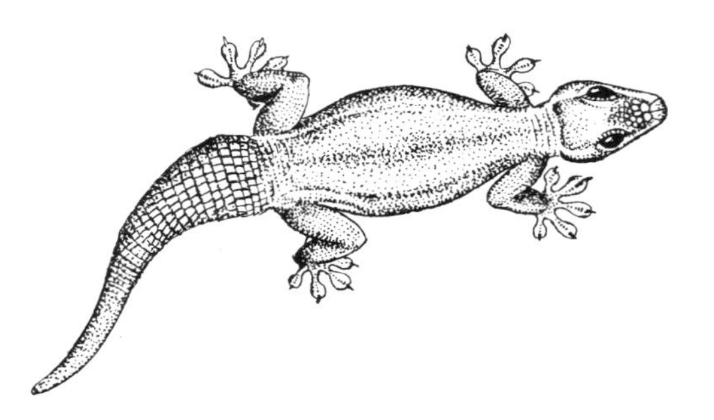

in wait by the palm trees. The *Relation de l'Île de Rodrigue* of 1709 also mentions the extinct Little Owls feeding upon these lizards.

Leguat stressed the geckos' tameness, though, and described them taking fruit from his hand. Their main diet seems to have been palm-fruit, flies and other insects. The Abbe Pingré, who observed the 1761 transit of Venus, described the Day Gecko on Rodriguez itself and on Booby Island and noticed the quick colour changes typical of this genus. Only six specimens exist of the Day Gecko but we have a good first-hand account by a naturalist from the nineteenth century. François Lienard, who lived on Mauritius, took an interest in Rodriguean fauna and had a small Day Gecko sent to him in 1842. He named it the 'Gecko à gorge jaune' and his descriptions are valuable because the colours of museum specimens fade so much.

By 1874, when a British expedition was on the island to record another transit of Venus, the Day Gecko was already confined to the islets; cats, rats and deforestation having wiped them out on Rodriguez. The last specimen ever seen was taken by M. Etienne Theroux from an islet in 1917 and extensive searches of all the satellites in 1963, 1967 and 1974 have been in vain.

RODRIGUEZ NIGHT GECKO

Gekkonidae
Phelsuma gigas
EXTINCT *c.* 1841
Rodriguez and satellites

The Rodriguez Night Gecko (*Phelsuma gigas*) was a much less colourful creature but it was massive for its genus, measuring up to 54cm (21in.) in overall length and with a girth of 20cm (8in.) or more. Greyish, mottled with black above, pale yellow below, and pink-tongued, it was described by Leguat as 'big and long as one's arm' and 'very ugly in appearance'. It was strictly nocturnal and hid by day among rocks or up Lantan palm trees. It was only known from Rodriguez and one islet .4km (.25 mile) offshore.

The Night Gecko was said to eat both fruit and meat, and to prey especially upon birds' eggs. However, when François Lienard had five specimens sent to him from the Ile aux Fregates in 1841, he found that they would only take sweetened water which he fed them from a spoon. His largest specimen was only 38cm (15in.) long and survived in captivity for just two months. Lienard's Night Geckos were the last ones ever recorded and the Isle aux Fregates now swarms with rats.

MAURITIAN GIANT SKINK

Scincidae
Didosaurus (Leiolopisma) mauritianus
EXTINCT *c.* 1650
Mauritius

Only guesswork can explain the extinction of another large Mascarene lizard, the Mauritian Giant Skink (*Didosaurus mauritianus*), for it is only known from subfossil bones. These were found in caves and in the Mare aux Songes swamp, whence most Dodo relics come, and the Mauritian Institute have managed to mount a complete skeleton.

Yet this skink seems to have been alive during the early years of European occupation and it is a mystery why no one ever noticed or recorded it. Hypothetically, we can suggest that it was nocturnal and hence inconspicuous despite its size (more than 50cm, 20in.), and that it, or its eggs, were destroyed by pigs and rats.

CAPE VERDE GIANT SKINK

Scincidae
Macroscincus coctei
EXTINCT *c.* 1940
Cape Verde Islands, E. Atlantic

The Mauritian Giant Skink was unique but was probably most closely allied to the Cape Verde Giant Skink (*Macroscincus coctei*), which is itself monotypic and also, almost certainly, extinct.

Within historical times the Giant Skink was confined to the two smallest islands in the 15-island Atlantic group: Branco (3sq km, 1.2sq miles) and Razo (7sq km, 2.7sq miles), both steep, uninhabited rocky outcrops from the ocean, with sparse vegetation, and considerable drifts of sand on their southern flanks.

The second largest of living skinks, the Cape Verde Giant shared its isolated habitat with a giant gecko and, in their breeding season, with a colony of Leach's Petrels. Often the skinks inhabited petrel nesting burrows, but this was not a case (as with the Tuatara) of an ectothermal reptile enjoying the company of a warm-blooded bird. For at this time of the year the normally vegetarian Skink apparently entered a carnivorous phase, and preyed upon the birds, their eggs and their young. The rest of the year the skink seems to have fed primarily upon the seeds of Mallows, ten scrubby species of which form the sole vegetation on Branco and Razo.

The last known specimens of the 'Lagaro', as it was called by the Cape Verde fishermen, were captives in German collections around 1914, but the islanders claim that it did not finally die out in the wild until about 1940. It had long been prized as a source of meat by visiting fishermen and suffered especially from human persecution in 1833 when, during a famine, a group of 30 convicts were put ashore on Branco and left to fend for themselves.

The main cause of the Giant Skink's extinction, however, seems to have been severe droughts, and consequent vegetation loss and soil erosion. Hans Hermann Schleich searched both islands in vain in 1979, though he is unwilling to confirm the extinction of the Lagaro categorically until he has searched a last, steep area of bare rock on Branco. The islanders told him emphatically that he had come 40 years too late.

Cape Verde Giant Skink *Macroscincus coctei*

MEDITERRANEAN LIZARDS

Lacertidae
RATAS ISLAND LIZARD
Podarcis lilfordi rodriquezi
EXTINCT *c.* 1950
Ratas Island Minorca
SAN STEPHANO LIZARD
Podarcis sicula sanctistephani
EXTINCT *c.* 1965
San Stephano Island Tyrrhenian Sea

In the Mediterranean, there have been two reptile extinctions in recent years, both of the genus *Podarcis*. The first, the Ratas Island Lizard (*Podarcis lilfordi rodriquezi*) became extinct about 1950 because its habitat – a small, rocky island in the bay of Mahon, Minorca – was entirely destroyed in the rebuilding of Port Mahon. The second, the San Stephano Lizard (*Podarcis sicula sanctistephani*) inhabited a similarly small island near Ventotene in the Tyrrhenian Sea. Its extinction about 1965 was probably caused by predation of feral cats and one species of snake, followed by an epidemic which seems to have destroyed virtually the whole population. Any survivors seem to have interbred with another, accidentally introduced subspecies.

WEST INDIAN LIZARDS

Iguanidae
JAMAICAN IGUANA
Cyclura collei
EXTINCT *c.* 1968
NAVASSA IGUANA
Cyclura cornuta onchiopsis
EXTINCT *c.* 1966
NAVASSA ISLAND LIZARD
Leiocephalus eremitus
EXTINCT *c.* 1900
MARTINIQUE LIZARD
Leiocephalus herminieri
EXTINCT *c.* 1837

Teiidae
MARTINIQUE GIANT AMEIVA
Ameiva major
EXTINCT *c.* 1960
GRAND ISLET (GUADELOUPE) AMEIVA
Ameiva cineracea
EXTINCT *c.* 1920

Anguidae
JAMAICAN GIANT GALLIWASP
Diploglossus occiduus
EXTINCT *c.* 1880
West Indies

The West Indies has suffered at least seven lizard extinctions in the past two centuries. During the 1960s two very striking species of iguana became extinct. The Jamaican Iguana (*Cyclura collei*) was a large species (113cm, 44.5in.) which became extinct because it was hunted for food and 'sport' by man, and its eggs and young were taken by the introduced mongoose. The other iguana, the Navassa Iguana (*Cyclura cornuta onchiopsis*) was found on the United States' West Indian island of Navassa. Its extinction was probably due to introduced rats, cats and goats. Navassa Island was also host to another lizard of the Iguanidae family, *Leiocephalus eremitus*, which became extinct about 1900, largely through predation by domestic cats. The related Martinique Lizard (*Leiocephalus herminieri*) disappeared even earlier. The reasons for its extinction have not been determined, but it was gone by 1837 at least.

Martinique was also the habitat for the Martinique Giant Ameiva (*Ameiva major*) which became extinct about 1960, possibly because of mongoose predation. The other extinct amieva lizard, the Grand Islet Ameiva (*Ameiva cineracea*) which inhabited an islet off Petit-Bourg on the east coast of Basse-Terre, Guadeloupe, became extinct about 1920 when the islet was devastated by a hurricane.

The large anguid lizard, the Jamaican Giant Galliwasp (*Diploglossus occiduus*) became extinct about a century ago, when the mongoose was making inroads into its territory. The last of these unusual lizards probably found refuge in the mongoose-free Hellshire Hills, an area believed to be the last retreat of the Jamaican Iguana, as well.

VEGAS VALLEY LEOPARD FROG

Ranidae
Rana pipiens (onca?) fisheri
EXTINCT *c.* 1966
Nevada, USA

Turning from true reptiles to amphibians there are two extinctions to record in the Frog family. The first of these, the Vegas Valley Leopard Frog (*Rana pipiens (onca?) fisheri*) was a relict population in Clark County, Nevada, depending for its habitat on freshwater springs and seepage areas. It differed externally from other Leopard Frogs in its very faint (or absent) spotting, and in the absence of the usual white jaw-stripe. It is the only amphibian known to have disappeared in North America in modern times though many other forms are threatened for similar reasons.

Las Vegas is known, of course, as the gambling capital of the world and – set as it is in the desert – its water demands are large and increasing. Thus a large part of the Leopard Frog's habitat has been removed since the 1930s simply by the capping and diverting of springs. In those small areas of open water that do survive there have been introductions of bullfrogs which eat their smaller cousins, and trout which eat the eggs and tadpoles. No Leopard Frogs have been seen in Vegas Valley since 1942 and related forms in other desert areas (such as Imperial County, California) will suffer the same fate unless firm conservation measures are taken.

PALESTINIAN PAINTED FROG

Discoglossidae
Discoglossus nigriventer
EXTINCT *c.* 1956
Hula Lake, Israel-Syria border

Twenty miles north of the Sea of Galilee and right on the war-torn Israeli-Syrian border lies Hula Lake. The swamps of its eastern shore harboured a relict species of frog, the Palestinian Painted Frog (*Discoglossus nigriventer*), which seems to have been obliterated through large-scale land-reclamation by the Israelis.

The Painted Frog was only discovered in 1940, when two immatures and two tadpoles were collected. It is the only member of its genus ever recorded from the east side of the Mediterranean and differs in many ways from the two other Painted Frogs, most noticeably in its much longer front legs. Growing to a length of at least 8cm ($3\frac{1}{4}$in.), the Painted Frog was ochre and rusty coloured above and greyish-black below, with numerous white dots around its glandular orifices.

The first two frogs collected were kept in a terrarium where the larger one (a female) swallowed its companion. It was not until the only other specimen was found, in 1955, that scientists realized that the first two had been immatures, for the new specimen, also a female, was twice as large. It was probably the last of its kind but fortunately its collector, Mr M. Costa, kept it alive for some time and learned something of its lifestyle. Active only at night, it spent the day burrowed into sand (excavated with its front legs) with its head just protruding from the water. This reflects a specialized adaptation to shallow swamp conditions which have now disappeared.

Palestinian Painted Frog *Discoglossus nigriventer*

New Zealand Grayling *Prototroctes oxyrhynchus* Extinct *c.* 1923

The Fish

SEA, RIVER AND LAKE FISH

As supplies for the human race become of increasingly critical concern, our awareness of the fragility of water systems is growing. Perhaps more than with any other section of our environment, the effects of industrialization and settlement on watersheds and reservoirs have been highlighted by media attention. We know the effects of deforestation and consequent erosion on watersheds: some Caribbean islands have lost all their water. We know the effects of atmospheric pollution on standing water: in November 1980 the Canadian government acknowledged that acid rain, drifting north-east from the US industrial cities, had 'killed' 30 lakes in Ontario. And we are beginning to appreciate the effects of direct water pollution as fish disappear from rivers and lakes or are found to contain mineral residues (especially mercury) which are dangerous to human consumers.

Additional hazards to the fish populations have come from western man's habit of bringing with him exotic species, usually game-fish from home, which have preyed upon native species. The introduction of British Brown and Rainbow Trout was almost certainly the main

NEW ZEALAND GRAYLING

Prototroctidae
Prototroctes oxyrhynchus
EXTINCT *c.* 1923
New Zealand

factor in eliminating the New Zealand Grayling (*Prototroctes oxyrhynchus*), itself a beautiful sporting fish, which has not been seen since 1923.

The grayling was superficially troutlike apart from its high dorsal fin and grew to a considerable size, some being recorded at 50cm (20in.) in length and weighing as much as 1.4kg (3lb). Its unusual and somewhat mysterious life-cycle resulted in specimens being caught of quite different colouration, for it almost certainly spent part of the year at sea and, when caught on its way upstream, was of a silvery hue with a slate-blue back. After several months in the rivers, however, the New Zealand Grayling was often a rich red brown, speckled with grey, and almost golden-hued on its belly.

The grayling moved in shoals and was reported at times in 'upstream migrations of huge shoals of ripe fish'. It was rarely found more than 48km (30 miles) from salt water, and preferred swift, stony-bedded streams and rivers. Though often caught on baited hooks, its teeth and intestines suggest that it primarily ate algae, and most contemporary reports speak of it as a night-feeder.

There are, however, few reliable descriptions of these fish, for although they were plentiful in swift waters on both islands when European settlers first moved in in the 1860s, they had virtually disappeared by the end of the century. When one was caught and brought to an hotel in Poptiki in 1904 it was exhibited as a great curiosity. Thus no scientists ever studied the grayling and most information is speculative or based on analogy with its only known relative (*P. maraena*) the 'cucumber herring' of southeastern Australia and Tasmania. It is even possible that the fish spawned at sea, since no specimens smaller than 12.5cm (5in.) were ever taken in fresh water.

The sea-going cycle of the New Zealand Grayling may explain some of the puzzles of its disappearance. For it vanished as rapidly from streams where trout had not been introduced and shade-trees had not been felled as from waters where they had. But if it had migrated to dangerously silted or overlit streams via the ocean, it could have suffered in that way. It should be remembered, too, that both forms of introduced trout also take to the sea and rapidly colonize neighbouring rivers where they have not been stocked.

In March 1923 Te Rangi Hiroa, otherwise known as Sir Peter Buck, caught a number of New Zealand Grayling in the Waiapu River in far eastern North Island. These were the last grayling reliably recorded, though rumours of their survival in the remote lower Westland district of South Island are still current.

THICKTAIL CHUB

Cyprinidae
Gila crassicauda
EXTINCT c. 1854
California

A not dissimilar fate befell a fish of the Central Valley in California, USA, though for the Thicktail Chub or Minnow (*Gila crassicauda*) there were the additional hazards of dam-building and introduced relatives. Interbreeding and hybridization is yet another consequence of human interference with native fish stocks.

Excavation in aboriginal Indian middens show that the Thicktail Chub was the third most abundant food source before European settlement. The chub, which reached a length of 30cm (12in.) (the females were larger), favoured the sluggish parts of streams, sloughs and ditches, and was evidently still numerous up till 1880 when it

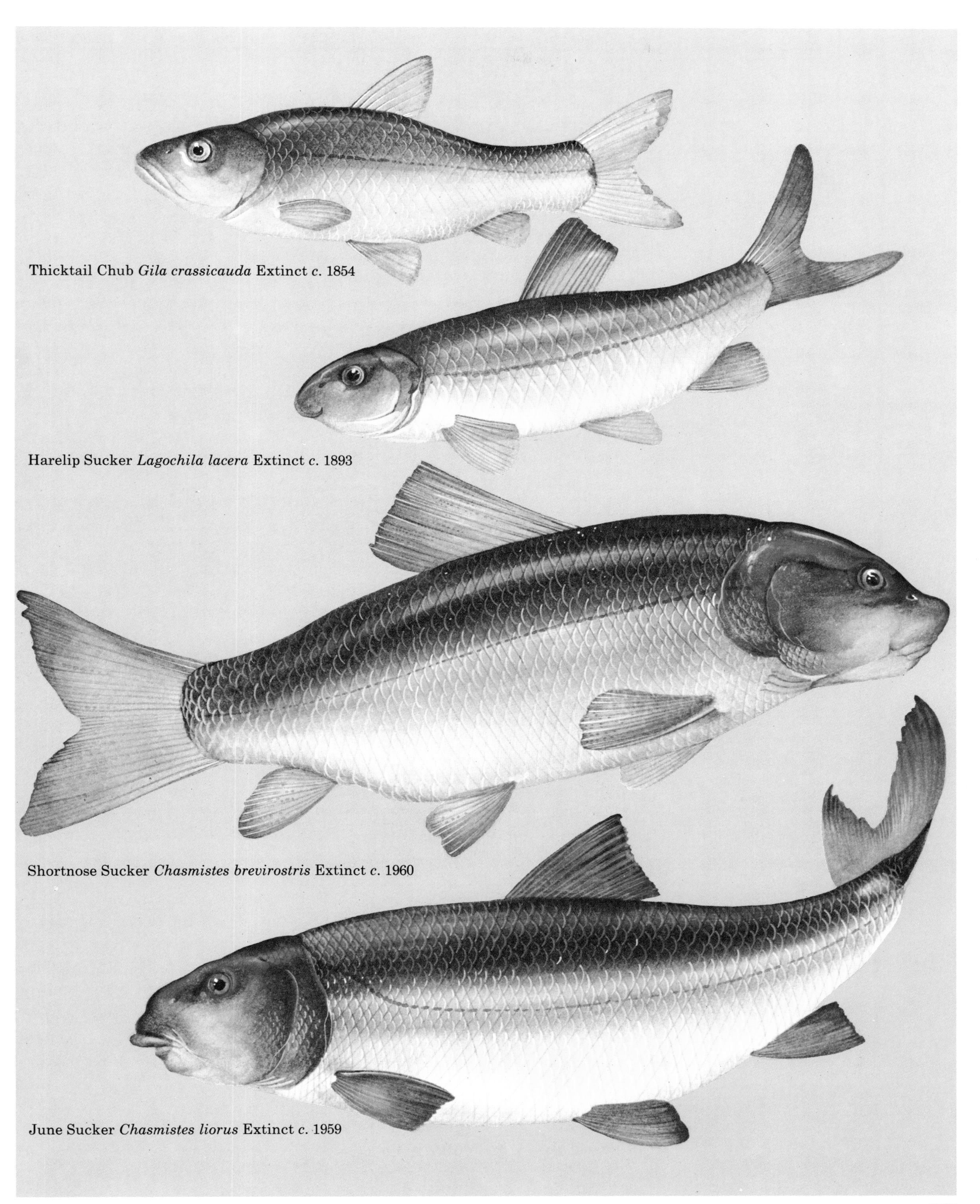

Thicktail Chub *Gila crassicauda* Extinct *c.* 1854

Harelip Sucker *Lagochila lacera* Extinct *c.* 1893

Shortnose Sucker *Chasmistes brevirostris* Extinct *c.* 1960

June Sucker *Chasmistes liorus* Extinct *c.* 1959

could still be bought in the San Francisco market. It was a short-headed, rather hump-backed fish, greenish-brown to purplish black above and yellowish below, and was most commonly taken from Coyote Creek which ran from Clear Lake into San Francisco Bay. At times of heavy spring flooding it was caught also in the surface waters of the bay itself.

Most of its habitat was destroyed by drainage, dam-building and water-diversion for irrigation in the first years of this century. By the late 1940s the Thicktail Chub had disappeared from Clear Lake too, and the last one recorded was taken from the area of the Sacramento River in Solano Country where it had first been described in 1854. But even in the 1920s hybrids between the Thicktail Chub and the introduced Hitch Minnow (*Lavinia exilicauda*) had been discovered and this was presumably the final factor in its disappearance as a pure-bred fish.

HARELIP SUCKER

Catostomidae
Lagochila lacera
EXTINCT *c.* 1893
Central USA

Yet another hazard for river fishes is the alteration to riverbeds caused by damming, change in the speed of currents, and above all by the erosion which follows hillside logging, where soil washed down by rain silts up gravel or stony bottoms. Just such a combination of human alterations spelt doom for the Harelip Sucker (*Lagochila lacera*), a highly adapted bottom-feeder which bore a superficial resemblance to the beautiful Redhorse Suckers which shared its range but were less specialized and have survived.

The Harelip Sucker, grew up to 45cm (18in.) long and to a weight of 1.4kg (3lb). Its back was grey to olive in colour and its tail slatey-blue, while its belly was whitish except in mineral-polluted waters where it took on a muddy-green cast. The Harelip Sucker was widespread in the basins of the Ohio, Maumee and lower Mississippi Rivers and was known by different names in various states. Thus the fishermen of the Columbus and Scioto Rivers (where its concentration was greatest) knew it as the 'May-Sucker' from its spawning runs in that month; while in the state of Indiana it was called the 'Pea-lip Sucker'.

That name gives some impression of the Harelip Sucker's highly specialized small mouth, downturned and proboscis-like with completely divided lower lip. This adaptation was ideal for riverbeds of clear limestone, gravel, sand or clay, but M. B. Trautman's monograph, *The Fishes of Ohio*, describes the change of environment graphically: 'Between 1850–1900 those portions of the Maumee and Scioto systems which originally were of the clear-water, prairie-stream type, have been converted into turbid streams of the western plains-type.' Not only did this remove the Harelip Suckers' food source but, in Trautmann's words, 'The Harelip, with its small, specialised mouth and closely bound gill-covers must have been particularly susceptible to asphyxiation' from the silt. The last Harelip Sucker was caught in 1893.

SHORTNOSE SUCKER

Catostomidae
Chasmistes brevirostris
EXTINCT *c.* 1960
Oregon

It seems that dam-building alone caused the extinction of another sucker, the Shortnose (*Chasmistes brevirostris*) which was primarily a lake fish but depended on rivers for its spawning run. Upper Klamath Lake in Oregon was its home and it spawned in Williamson and Sprague Rivers. It was once the chief subsistence food of the

Klamath Indians who, incidentally, denied that it entered the rivers, but unless some so-far-undetected ecological change has occurred in the lake itself, the river dam-barriers seem to be the only possible cause of its decline and disappearance. The slump in numbers was noticed during the last war, and no specimen of the Shortnose Sucker has been taken since 1960.

JUNE SUCKER
Catostomidae
Chasmistes liorus
EXTINCT *c.* 1959
Utah

The closely related June Sucker (*Chasmistes liorus*) was, until 1935, the staple of a thriving commercial fishery in Utah Lake, Utah; but the fishery was abandoned in that year after severe droughts and increased domestic use of its spawning river, the Provo, reduced its numbers catastrophically. The next year, for the first time in history, there was no spawning run in the spring and there has not been another spawning since. The only fish caught thereafter seem to have been isolated survivors, and these have usually been hybrids with related species. The last of these were caught in the Provo River in 1942 and in Utah Lake in 1959.

SPRING VALLEY SUCKER
Catostomidae
Pantosteus sp.
EXTINCT *c.* 1950
Nevada

A much smaller member of the Sucker family, endemic to a single small stream in eastern Nevada, seems to have disappeared within the last thirty years. This was the Spring Valley Sucker (*Pantosteus sp.*), known only from one specimen collected in 1938 and never fully described by scientists. (For this reason it is not illustrated here.)

The Spring Valley Sucker was found at the northern end of Spring Valley but there was no sign of it when, in 1959, Drs R. R. Miller and Carl Hubbs worked thoroughly along its home stream, and searched adjacent waters too. They found no fish at all in the valley-bottom streams, with the exception of one minnow species; while the mountain streams running down to the valley contained only recently introduced species of trout. One might surmise that the trout had preyed upon the Spring Valley Sucker or its eggs, but Dr Miller's theory is that the endemic fish more likely suffered from a deterioration in the stream's flow.

A large proportion of the extinctions described in this book occurred on islands, and we should not forget that isolated bodies of water are themselves islands of a kind. Lake fishes, then, are as likely as other animals to adapt over the centuries to peculiar ecological niches and then to succumb quickly to the harrassments of Europeans and the creatures they bring with them.

As with islands, the larger a lake the more chance its endemic fauna has of surviving change and, as with mountainous islands, deep lakes are more likely to provide remote havens for its native animals to retreat to. Yet the Great Lakes of North America, virtually inland freshwater seas, have suffered several fish casualties and are affected by every conceivable human pressure and pollutant from massive industrial development, and the establishment of continually growing city populations on both the American and Canadian shores. The great depth of these lakes has proved little protection, either, in the face of invasion by an exotic marine fish, the Sea Lamprey (*Petromyzon marinus*) which may partly have established itself by migrating up the Saint Lawrence Seaway, but was more probably introduced by ships using the Seaway and flushing their bilges into the waters of Lake Michigan and Lake Huron.

BLACKFIN CISCO
DEEPWATER CISCO

Coregonidae (Leucyths)
Coregonus nigripinnus
Coregonus johannae
EXTINCT *c.* 1960
Lakes Michegan and Huron, USA and Canada

The Blackfin Cisco (*Coregonus nigripinnus*) and the Deepwater Cisco (*Coregonus johannae*) were abundant in both these lakes well into this century, but have now disappeared, while their relatives in Lake Erie have virtually gone. These ciscos are quite unusual for their family, perhaps from long isolation in their lacustrine 'islands', and many ichthyologists prefer to place them in a separate family (Leucicthys).

For the fishermen on the Great Lakes in the last century, though, the ciscos were 'Jumbo Herring', abundant and available summer and winter and extremely lucrative. The pressures of fishing, by gillnet and by set 'traps' between the islands, were intensive and sustained, and by World War I the numbers of cisco had noticeably declined. Since the fisherman had no conception of conservation or of an off-season, this is hardly surprising. When the fish 'ran' in November huge catches were taken – one lake alone had yielded over 15 million tons (14.6 million tonnes) in 1885; it was common for one net to take out as much as ten tons (9.8 tonnes) in a day; and this glut of fish all at one time before freezing facilities were available led to widescale dumping of unsold catches. Even in mid winter, when the fish were lying very deep, the fishing continued. Holes were cut in the ice and long lines dropped through, and there are records of one man ice-fishing with a pearl-button for bait and hauling up 136kg (300lb) of cisco in a day.

The 'Jumbo Herring' formed the most important commercial catch in the Great Lakes and were completely 'fished out'. The end of commercial fishing will usually give surviving fish a chance to recover and build up reasonable stocks again, but the aforementioned Sea Lamprey seems to have prevented any hope of this. Proliferating and gorging like the rats on the Mascarenes or the mongooses in the Caribbean islands, the Sea Lampreys have apparently wiped these ciscos out completely. Neither the Blackfin Cisco nor the Deepwater Cisco has been recorded since 1960.

LAKE TITICACA ORESTIAS

Cyprinidontidae
Orestias cuvieri
EXTINCT *c.* 1950
Peru-Bolivian Border

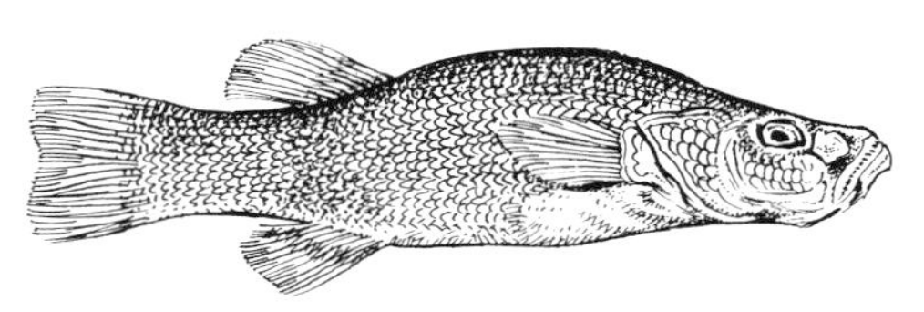

Lake Titicaca Orestias
Orestias cuvieri

Lake Titicaca, 4000m (13,000ft) up in the Andes on the border between Peru and Bolivia, is a pygmy compared to the North American Great Lakes but it is still, with a length of 240km (150 miles), a massive stretch of water and it is very deep. Indeed, ichthyologists have divided it into 'ecozones' by depth, each sustaining different species, and they have yet to investigate those zones deeper than 40m (130ft).

The most characteristic fishes of this and a few neighbouring lakes and rivers are the Orestiinae: small, cold-water fish that can be separated into at least 20 species. The largest of these (*Orestias cuvieri*) appears to have been destroyed, and many others are at risk because of the quite senseless introduction in 1937 of Lake Trout (*Salvelinus namaycush*) by a body which should have known better, the US Fish and Wildlife Service.

This Orestias was first collected in the 1830s and presented a peculiar sight indeed, with a mouth so upturned as to face almost vertically and a consequently concave head, which took up a full third of the overall body length. Full grown adults could measure

26.5cm ($10\frac{1}{2}$in.) and were known locally as 'Umautos', while the young were called 'Peje-Rey'. The adults were greenish-yellow to umber above, with a black lower jaw and black-striped fins. The scales were unusual in being very light at their centre. The 'Peje-Reys' were blotched and spotted, and are said to have congregated in deep, rock-bottomed zones in the cold season.

While other members of its family had established themselves in the Bolivian lakes, Poopo and Junim, and in three rivers and connecting waterways, the deepwater *Orestias cuvieri* was inhibited from migrating *via* shallower waters, and was thus confined to Lake Titicaca itself. Its ecozone was at about the 30m (100ft) mark, exactly the depth where the introduced Lake Trout would be most abundant. The local people caught the Umautos during seasonal 'migrations' in the Lake and these have not occurred since the last war. Selective netting in 1960 failed to procure a single specimen, though all the other Orestias – and numerous Lake Trout – were present in the haul.

UTAH LAKE SCULPIN
Cottidae
Cottus echinatus
EXTINCT *c.* 1936
Utah

One lake fish whose extinction cannot be blamed on human interference was the peculiar freshwater Sculpin (*Cottus echinatus*) of Utah Lake. Its disappearance seems to have been directly consequent on the droughts of the mid-1930s, mentioned above in connection with the June Sucker. If the drought and the fall in water levels did not totally destroy the Sculpin's spawning grounds they must have left its stocks very much depleted and it could well be that any survivors or young were killed off by other fish in the lake, including a number of established exotics.

Blackfin Cisco *Coregonus nigripinnus* (Above) Deepwater Cisco *Coregonus johannae* (Below)

Utah Lake Sculpin *Cottus echinatus* Extinct *c.* 1936

Tecopa Pupfish *Cyprinodon nevadensis calidae* Extinct *c.* 1960?

Ash Meadows Killifish *Empetrichthys merriami* Extinct *c.* 1948

Big Spring Spinedace *Lepidomeda mollispinnis pratensis* Extinct *c.* 1950

Parras Roundnose Minnow *Dionda episcopa plunctifer* Extinct *c.* 1930

The southwestern United States contain a number of drainage systems, relics of a huge Pleistocene watershed, where isolated pockets of water, often quite tiny, have supported small populations of little fish until this century. There have been many extinctions here – all caused by human interference – which have been documented largely by Robert Rush Miller of the University of Michigan and through the fieldwork of Carl and Laura Hubbs. Dr Miller has highlighted the effects of shrinking water flows and the obstructions caused by barrier dams, but has emphasized particularly the plight of the many species which have limited ranges: 'some are even confined to habitats that contain only a few thousand gallons of water or scarcely cover an acre.'

STUMPTOOTH MINNOW

Cyprinidae
Stypodon signifer
EXTINCT *c.* 1930
Chihuahua, Mexico

PARRAS ROUNDNOSE MINNOW

Cyprinidae
Dionda episcopa plunctifer
EXTINCT *c.* 1930
Chihuahua, Mexico

PARRAS PUPFISH

Cyprinidontidae
Cyprinodon latifasciatus
EXTINCT *c.* 1930
Chihuahua Mexico

In this category fall three species once found in the Chihuahua Desert, near Parras, Coahuila, Mexico. The Stumptooth Minnow (*Stypodon signifer*), the Parras Roundnose Minnow (*Dionda episcopa plunctifer*) and the Parras Pupfish (*Cyprinodon latifasciatus*) have all disappeared since the 1930s when the related springs which supported them were diverted by man. The Parras springs emerged from a lava hillside and are relics of the old Rio Nazas Watershed, and their meagre flow has been combined by tunnels cut in the lava, and channelled into a reservoir which feeds first a cotton mill and then a series of irrigation ditches. To compound the effects of this alteration the reservoir has been stocked with carp which would quickly have swallowed any surviving small fish.

The Stumptooth formed a unique genus and species and is only known from six specimens, the last four taken in 1903 by George Hochderfer. He took nine Pupfish at the same time from a neighbouring spring, also the last record of that fish. The Roundnose Minnow was found in a small spring in Saltillo as well as Parras but seems to have disappeared from both localities. The Hubbs' investigations in 1953 found that 'there no longer are any natural springs in the immediate vicinity of Parras or anywhere in the Parras basin or in the great desert basin to the North'. The habitat of these species, then, has been totally destroyed.

ASH MEADOWS KILLIFISH

Cyprinodontidae
Empetrichthys merriami
EXTINCT *c.* 1948
Nevada

The Death Valley basin further north presents a similar picture. There the streams are relics of a water system which once reached as far as the Colorado River, and there are a number of highly individuated species, many at risk and some extinct. The Ash Meadows Killifish (*Empetrichthys merriami*) is one of two species of a genus confined to this area (Nye Valley, Nevada) and seems to have fallen prey to introduce crayfish and bullfrogs.

Structurally quite distinct from other cyprinodonts, it was a strange-looking little fish – deep-bodied, hump-backed and narrow-mouthed, with a markedly protruding lower jaw. Its sole habitat was 'Deep Spring', at 9m (30ft) easily the deepest pool in Ash Meadows. The water is described as 'a clear, chalky-blue', with a distinctly sulphurous odour. The bottom is formed of accumulated silt and mud, and the pool supports a dense growth of algae – the main food source of the Killifish and of the other cypridonts which shared the pool but favoured a higher level, above a rock-shelf on the pool's southern wall. The temperature of Deep Spring is about 20°C (68°F).

Separated from Ash Meadows by a low, alluvial divide, the other members of this genus, the Pahrump Killifish (*E. latos*) inhabited three slightly warmer (24°C, 75°F) springs. The nominate species survives in Pahrump Valley in tiny numbers, but two subspecies there (*E. latos pahrump* and *E. latos concavus*) have disappeared following the introduction of carp.

PAHRUMP KILLIFISH

Cyprinodontidae
Empetrichthys latos pahrump
Empetrichthys latos concavus
EXTINCT *c.* 1950
Nevada

The South Tecopa Hot Spring in Inyou County, California, harboured a pupfish (*Cyprinodon nevadensis calidae*) which was remarkable for thriving in the highest water temperature ever recorded for a fish (up to 40°C, 104°F). The lower, cooler stretches of this spring (which rapidly drains into marshland) were stocked with Mosquito Fish (*Gambusia affinis*), an exotic which has been particularly destructive in the southwest. This, and contamination from a bathhouse, seem to have eliminated the Tecopa Pupfish, which was only discovered in 1942, though some searchers have not yet given up hope.

TECOPA PUPFISH

Cyrinodontidae
Cyprinodon nevadensis calidae
EXTINCT *c.* 1960?
California

Introduction of the Mosquito Fish, however, was certainly the chief reason for the elimination of the Big Spring Spinedace (*Lepidomeda mollispinis pratensis*), an endemic of a single spring-fed marsh near Panaca, Lincoln County, Nevada, sometime between 1938 and 1959.

Another spinedace (*Lepidomeda altivelis*) from Ash Spring and a chain of lakes in Pahranagat Valley, Nevada, also suffered from predation by Mosquito Fish and other exotics, as well as habitat modification; while introduced Rainbow Trout and Brook Trout eliminated the Grass Valley Speckled Dace (*Rhinichthys osculus reliquus*) from its enclosed waterbasin in Lander County nearby, between 1938 and 1959.

BIG SPRING SPINEDACE
PAHRANAGAT SPINEDACE
GRASS VALLEY SPECKLED DACE

Cyprinidae
Lepidomeda mollispinis pratensis
Lepidomeda altivelis
Rhinichthys osculus reliquus
EXTINCT *c.* 1950
Nevada

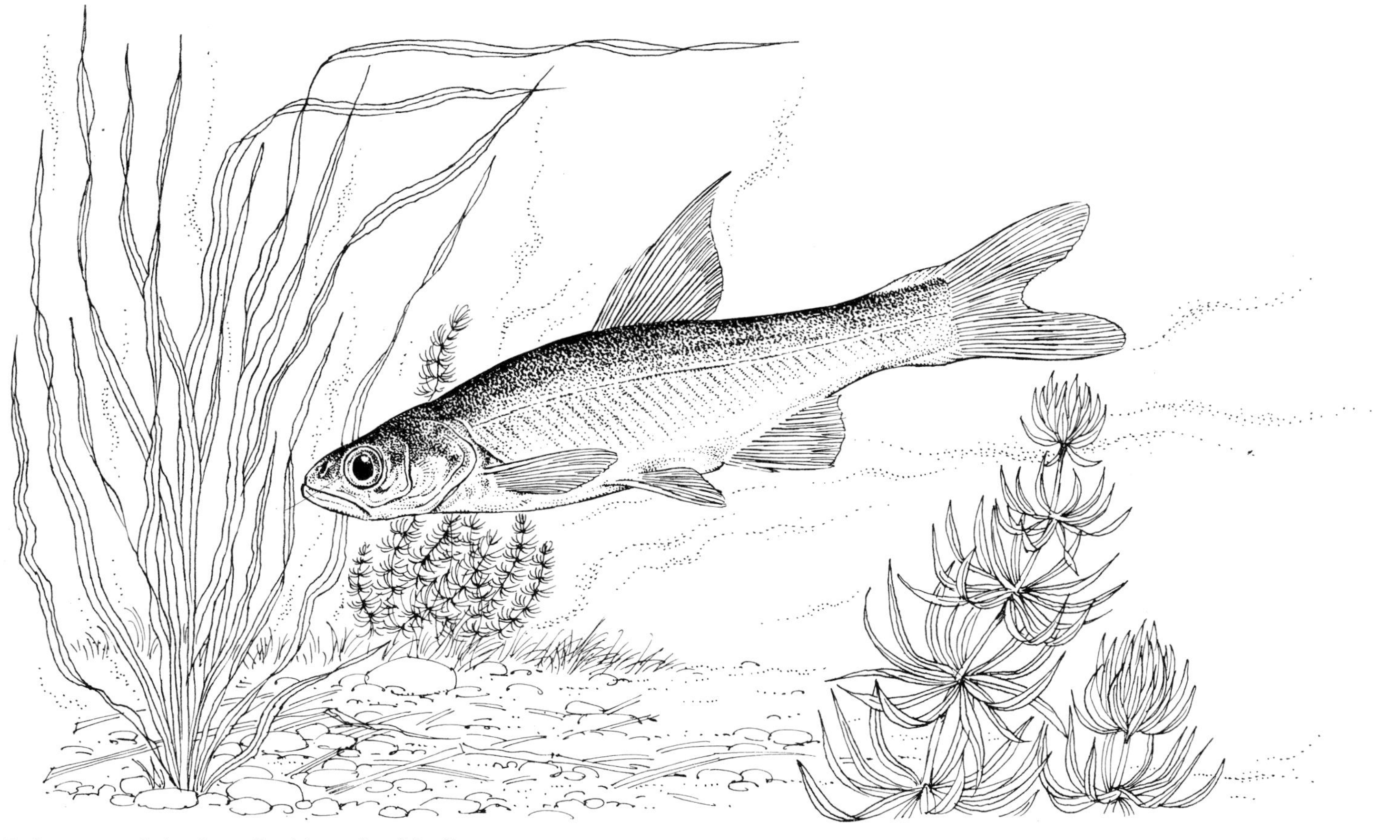

Pahranagat Spinedace *Lepidomeda altivelis*

Part Four

APPENDIX

Plants

FROM THE ORCHID TO THE DODO TREE

No one knows how many plant extinctions there have been over the three centuries with which this book is concerned. However, it is estimated that there were several thousand, and that today about 20,000 vascular plant species are endangered.

As with the vertebrate animals, the plants of certain geographic areas are more vulnerable than those of other regions. Isolated islands often have a high number of endemic varieties of plant life. Of the plants of Madagascar, for instance, 80 per cent are endemic. The Hawaiian Islands have 97 per cent endemism, of which about 300 species have become extinct and another 800 are endangered. Even more extreme extinction rates may be found on very small islands. On Napoleon's tiny south Atlantic island of exile, St Helena, it is estimated there were over 100 endemic species of plants – all but 20 are now extinct, and of these only 5 are not critically endangered.

Other areas with high rates of endemic and localized plant populations – which are therefore often highly vulnerable to extinction – are alpine regions, deserts and rainforests. Tropical rainforests display the highest plant diversities of any habitat, and the rainforests of that part of South America drained by the Amazon and its many tributaries (the scene of perhaps the greatest forest destruction in the world today) have the greatest plant diversity of any region in the world. Here, each square mile of forest may harbour as many as 600 plant species.

Obviously the scale of plant extinction is even greater than that of the higher animals' and finally, perhaps, even more dangerous, as virtually all life-forms are directly or indirectly dependent on plants or their by-products. Furthermore, it has often been pointed out that virtually the whole of man's agriculture is based on hybrid forms of only 30 natural species. It is a great waste of resources to destroy thousands of possibly valuable species before their potential is even known.

One category of extinction is extremely ironic, in that it is being created by those who obviously value aesthetically the species they are actively destroying. Large numbers of rare plants have become extinct entirely because of the beauty of their flowers. Commercial exploitation of rare flowering plants has become a worldwide industry. Probably the most seriously affected are the thousands of Orchid species which are rapidly being exterminated in the jungles of northern India, Brazil, Southeast Asia and Hawaii.

Among the other flowering plants seriously threatened by excessive collecting is the 'prima donna' of Sumatra's plants, the *Rafflesia arnoldii.* This rare and brilliantly-coloured parasitic plant produces the world's largest flower, measuring 100cm (3ft) or more in diameter. Similarly, the famous Venus Fly-trap *(Dionaea muscipula)* of North and South Carolina – which Darwin described as 'one of the most wonderful (plants) in the world' – and the equally insectivorous Canebrake Pitcher Plant *(Sarracenia alabamensis)* of Alabama, although legally protected, have been endangered in the wild by collectors.

Most plant extinctions, however, result from large-scale habitat destruction. Extinction by this means is usually unintentional, but the scale of ecological change modern man inflicts on the land often eliminates plant species that were once extremely common.

Extensive changes in irrigation systems on the Nile, for example, resulted in the almost total elimination of ancient Egypt's two most famous plants: the Nile Lotus *(Nymphaea lotus)* – symbol of the Kingdom of Upper Egypt; and the Egyptian Papyrus *(Cyperus papyrus hadidii)* – symbol of the Kingdom of Lower Egypt. In the case of the Egyptian Papyrus, this was remarkable as it was the most widely used plant in that country. Everything from paper, food and medicine, to mats, sandles, baskets, garlands, shelters and boats were made from this reed. In this century it was believed to be entirely extinct until 1968 when a tiny population was rediscovered in a small, undisturbed waterway.

Kawaihae
Hibiscadelphus bombycinas

One of the most fascinating areas of investigation in natural history is that of symbiotic relationships between plants and animals, which suggest theories of parallel evolution. Many of these cases involve rare and extinct plant and animal species, and emphasize most vividly the interdependence of all life forms.

In Hawaii, for instance, the Hau Kuahiwi *(Hibiscadelphus wilderianus)* flowering tree became extinct about 1912. A major factor in this large tree's disappearance seems to have been the rarity and final extinction of the nectar and insect feeding Honeycreepers *(Drepanididae)*. These beautiful birds served the trees in much the same manner as bees serve other flowering plants in the process of pollination. The coming of Europeans marked the beginning of their mutual downward spiral to extinction. Of 39 species and subspecies of Honeycreeper, 16 are extinct and 14 endangered. Of the 5 species of the *Hibiscadelphus* genus, 2 are extinct (*H. wilderianus* and *H. bombycinas)* and the other 3 are critically endangered. A similar state has also been reached (for similar reasons) by the Hawaiian Hawane Palms *(Pritchardiae)*.

One of the more mysterious symbiotic relationships between plant and animal involved the extinct Dodo and the tree *Calvaria major* which is endemic to Mauritius and survives as a species in only 13 living specimens. The fruit of this tree was known to be a major food supply for the Dodo, and so the *Calvaria* commonly became called the Dodo Tree, but recent investigations have proved this name to be doubly appropriate.

Although records show that the Dodo Tree was common on the island when Europeans first arrived, there is no evidence that a single new tree has appeared since the extinction of the Dodo in 1680. Although fruit and seed continue to grow on the 13 remaining 300 to 400 year old trees, no sapling has ever been produced, and the trees are nearing the end of their life span.

In 1973 the American, Dr Stanley Temple, investigated this puzzle while on Mauritius and in a series of experiments has come to the

Dodo Tree
Calvaria major

Mauritian Ebony
Diospyros

Wine Palm
Pseudophoenix ekmanii

convincing conclusion that only through the ingestion of the fruit and the crushing of its large seed-casing in the Dodo's powerful gizzard, could the seed germinate. By force-feeding the fruit to turkeys, Dr Temple managed to germinate three seeds – the first for 300 years. It is now thought that some system of artificial abrasion of the Dodo Tree's seeds may save it from extinction.

No such hope can be held out for many other plant species on Mauritius. Particularly regrettable is the loss of the island's *Diospyros* – a robust flowering hardwood tree. It is one of the larger true ebony trees, a potential source of commercial black ebony. However, like all of its genus, it is 'dioecious' – that is, its male and female flowers are borne on separate trees – and only one known *Diospyros* now exists. This is a 100 year old female. (A younger tree could be propagated by grafting.) With no male tree to 'mate' with, this ebony tree is effectively extinct.

Many other island or jungle species of trees of considerable commercial value have become extinct. A typical case was the exceptionally beautiful, red-grained and fragrant Juan Fernandez Sandlewood *(Santalum fernandezianum)* from the Isla Robinson Crusoe in the Juan Fernandez Island Group in the Pacific. This species was clear-cut and shipped to Peru. The island was then populated with goats which ate any surviving seedlings. This Sandalwood seems to have become extinct about 1916.

A more unusual case was that of the Wine Palm *(Pseudophoenix ekmanii)* of the Dominican Republic whose pleasant juicy sap, when fermented, made a fine, clear wine. The tree also produced a cherry-like fruit. Unfortunately, a kind of garrotting process was used to gather the sap, and this invariably killed the tree. Through over-exploitation, the Wine Palm became extinct about 1926.

Many plants are of extreme interest to science, as unique evolutionary experiments, such as the already mentioned insectivorous plants, or as 'living fossils' – survivors from a pre-historic dawn. In the animal world the endangered 'dinosaur' survivor, the Tuatara lizard, and the rare 300 million year old 'walking fish' called the coelacanth are examples of living fossils. Among plants we have such ancient survivors as the Maiden Hair Tree (*Ginko biloba*) and the Monkey Puzzle Tree (*Metasequoia glyptostrobioides*). Both now exist almost exclusively as cultivated plants.

The International Union for Conservation of Nature and Natural Resources (IUCN) *Red Data Book* of plants, in its short-listing of endangered species, names many that would prove directly useful, not just as timber, fuel, or to horticulture, but as potential food crops. They include a variety of olive *(Olea laperrinei),* a maize *(Euchlaena perennis),* an avocado *(Persea theobromifolia),* a date palm *(Phoenix theophrasti),* the high-protein, edible Yeheb Nut *(Cordeauxia edulis),* a pomegranate *(Punica protopunica),* and the Wild Rubarb *(Rheum rhaponticum)* – the nearly extinct breeding stock from which the cultivated rhubarb was derived during the seventeenth century.

The chosen area of concern and investigation for this book has been the vertebrate animals. However, the other far more numerous life forms: the vascular and non-vascular plants, as well as the invertebrate animals are as important – indeed, without them, no vertebrate could survive.

There are less than 50,000 vertebrate species – 60 per cent of which are fish. About 8,600 are birds, 7,700 are reptiles and amphibians, and 4,200 are mammals. (Among the mammals, if rats, bats, and insectivores are excluded, there are less than 1200 remaining 'higher' mammal species.)

In contrast, there are 5 to 10 million plants and invertebrates on earth. Estimates of the number of extinctions among these forms by the end of this century are 450,000 to 1,800,000. It is, of course, nearly impossible to relate to such numbers except in an abstract way and with the realization that such a pace of destruction is phenomenal. Obviously, we must attempt to reappraise our ideas of land exploitation to minimize the damage. Acknowledging this, we also come to see that the extinction of vertebrate animals is only the most immediate aspect of our alarming destruction of the earth's resources.

However, this short general essay on plant extinction was written to draw attention to some degree to the dilemma of all life forms. Plant extinction is obviously a field with vital concerns specifically its own. This essay can make no real inroad into such an immense subject. Its concerns can only be touched on here to draw attention to its critical state, and to demonstrate how the particular concerns of the vertebrate animals are not isolated. Indeed, there is no real possibility of even slowing down the extinction rate of the larger animals, if their ecological context is not taken into account in conservation schemes. There is little use in granting an animal legal protection, if licence is given to destroy its habitat. It must be recognized that a larger view of entire ecological systems must invariably be taken to ensure the protection of any one species.

Moth Orchid *Phalaenopis stuartiana*

Atlas of Extinct Species

1680 TO 1980

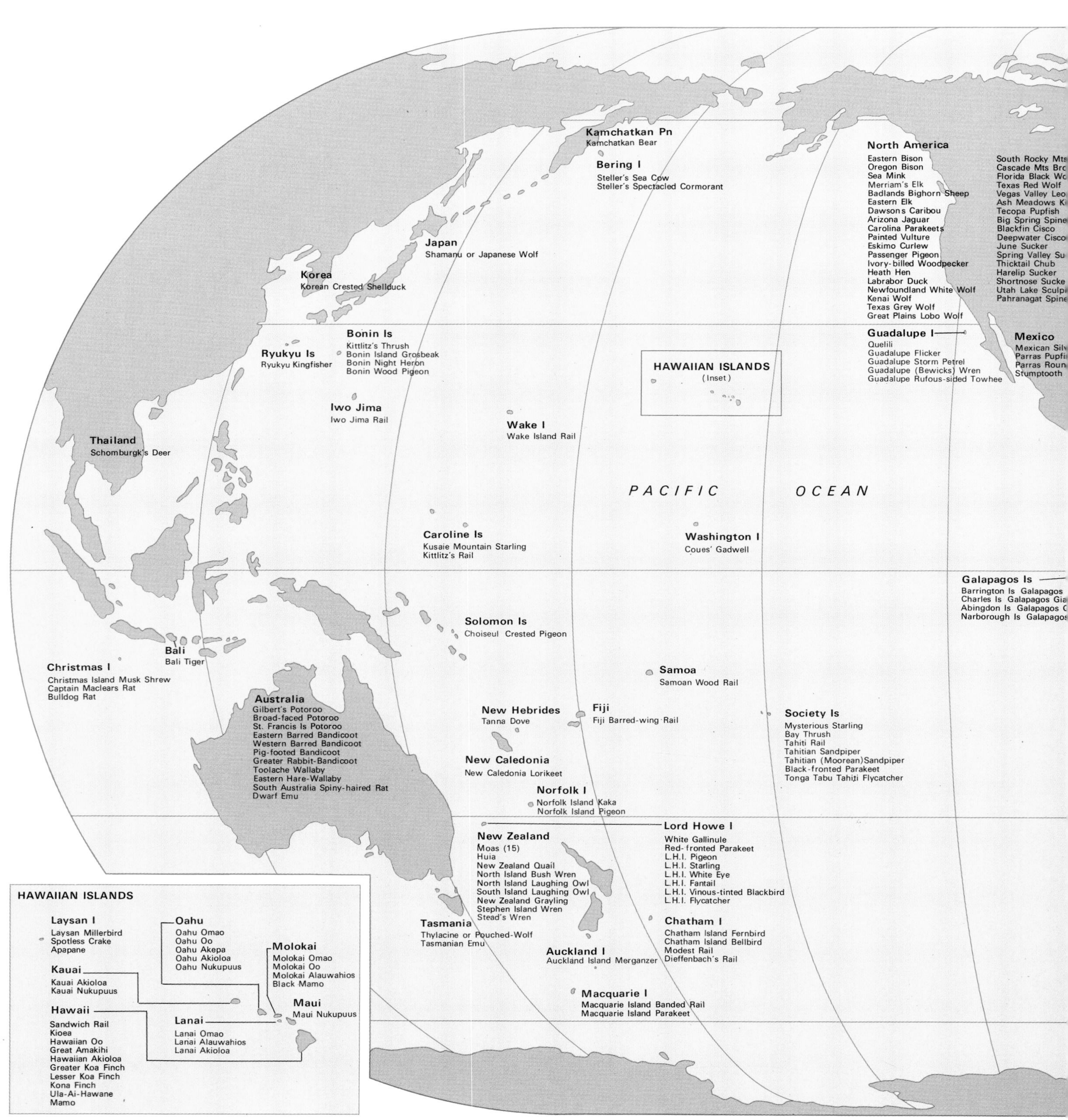

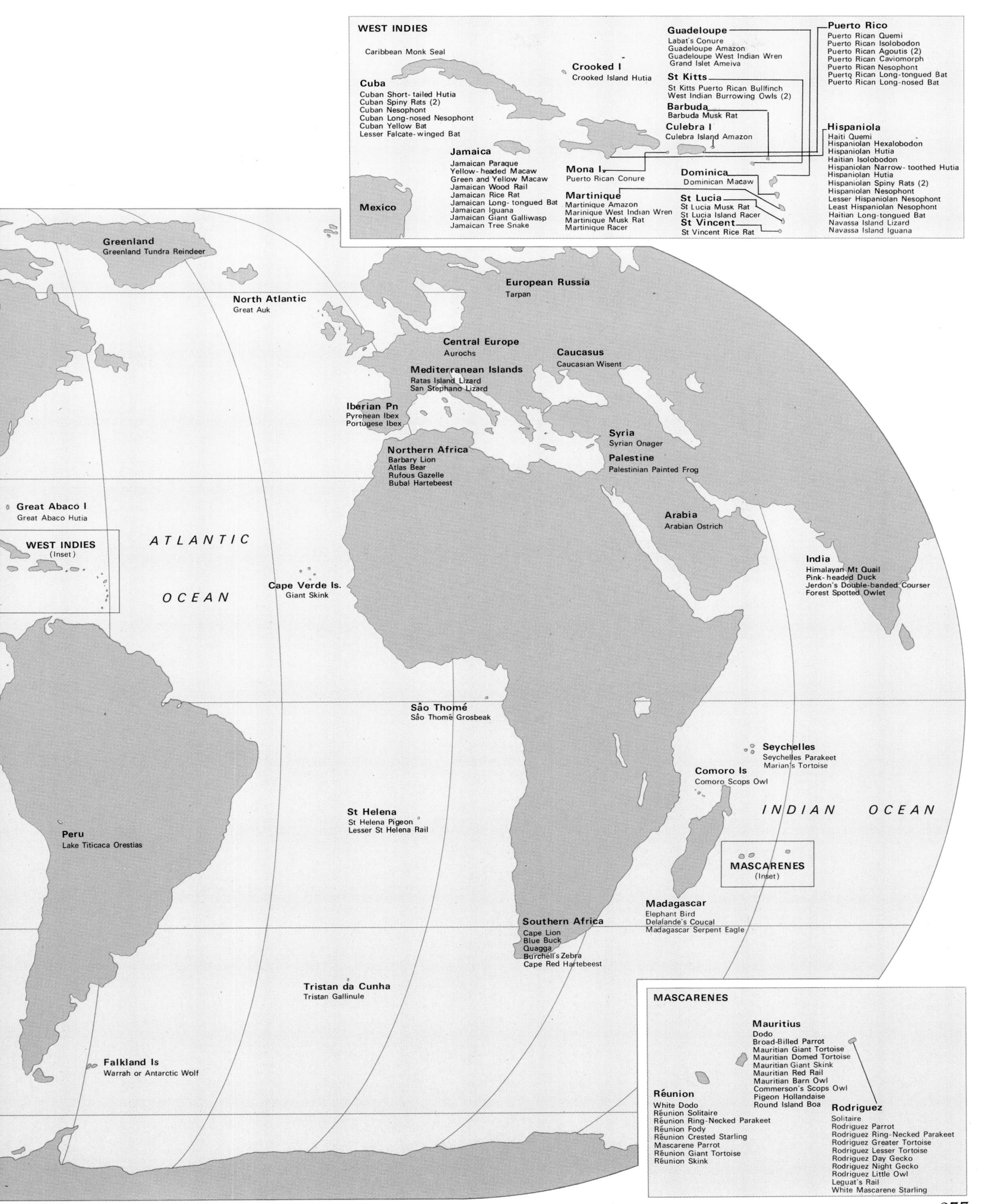
WEST INDIES
Caribbean Monk Seal
Cuba
Cuban Short-tailed Hutia
Cuban Spiny Rats (2)
Cuban Nesophont
Cuban Long-nosed Nesophont
Cuban Yellow Bat
Lesser Falcate-winged Bat
Jamaica
Jamaican Paraque
Yellow-headed Macaw
Green and Yellow Macaw
Jamaican Wood Rail
Jamaican Rice Rat
Jamaican Long-tongued Bat
Jamaican Iguana
Jamaican Giant Galliwasp
Jamaican Tree Snake
Mexico
Crooked I
Crooked Island Hutia
Mona I
Puerto Rican Conure
Martinique
Martinique Amazon
Marinique West Indian Wren
Martinique Musk Rat
Martinique Racer
Guadeloupe
Labat's Conure
Guadeloupe Amazon
Guadeloupe West Indian Wren
Grand Islet Ameiva
St Kitts
St Kitts Puerto Rican Bullfinch
West Indian Burrowing Owls (2)
Barbuda
Barbuda Musk Rat
Culebra I
Culebra Island Amazon
Dominica
Dominican Macaw
St Lucia
St Lucia Musk Rat
St Lucia Island Racer
St Vincent
St Vincent Rice Rat
Puerto Rico
Puerto Rican Quemi
Puerto Rican Isolobodon
Puerto Rican Agoutis (2)
Puerto Rican Caviomorph
Puerto Rican Nesophont
Puerto Rican Long-tongued Bat
Puerto Rican Long-nosed Bat
Hispaniola
Haiti Quemi
Hispaniolan Hexalobodon
Hispaniolan Hutia
Haitian Isolobodon
Hispaniolan Narrow-toothed Hutia
Hispaniolan Hutia
Hispaniolan Spiny Rats (2)
Hispaniolan Nesophont
Lesser Hispaniolan Nesophont
Least Hispaniolan Nesophont
Haitian Long-tongued Bat
Navassa Island Lizard
Navassa Island Iguana
Greenland
Greenland Tundra Reindeer
North Atlantic
Great Auk
European Russia
Tarpan
Central Europe
Aurochs
Caucasus
Caucasian Wisent
Mediterranean Islands
Ratas Island Lizard
San Stephano Lizard
Iberian Pn
Pyrenean Ibex
Portugese Ibex
Syria
Syrian Onager
Palestine
Palestinian Painted Frog
Northern Africa
Barbary Lion
Atlas Bear
Rufous Gazelle
Bubal Hartebeest
Great Abaco I
Great Abaco Hutia
WEST INDIES
(Inset)
ATLANTIC
OCEAN
Cape Verde Is.
Giant Skink
Arabia
Arabian Ostrich
India
Himalayan Mt Quail
Pink-headed Duck
Jerdon's Double-banded Courser
Forest Spotted Owlet
São Thomé
São Thomé Grosbeak
Seychelles
Seychelles Parakeet
Marian's Tortoise
Comoro Is
Comoro Scops Owl
St Helena
St Helena Pigeon
Lesser St Helena Rail
INDIAN OCEAN
Peru
Lake Titicaca Orestias
MASCARENES
(Inset)
Madagascar
Elephant Bird
Delalande's Coucal
Madagascar Serpent Eagle
Southern Africa
Cape Lion
Blue Buck
Quagga
Burchell's Zebra
Cape Red Hartebeest
Tristan da Cunha
Tristan Gallinule
Falkland Is
Warrah or Antarctic Wolf
MASCARENES
Mauritius
Dodo
Broad-Billed Parrot
Mauritian Giant Tortoise
Mauritian Domed Tortoise
Mauritian Giant Skink
Mauritian Red Rail
Mauritian Barn Owl
Commerson's Scops Owl
Pigeon Hollandaise
Round Island Boa
Réunion
White Dodo
Réunion Solitaire
Réunion Ring-Necked Parakeet
Réunion Fody
Réunion Crested Starling
Mascarene Parrot
Réunion Giant Tortoise
Réunion Skink
Rodriguez
Solitaire
Rodriguez Parrot
Rodriguez Ring-Necked Parakeet
Rodriguez Greater Tortoise
Rodriguez Lesser Tortoise
Rodriguez Day Gecko
Rodriguez Night Gecko
Rodriguez Little Owl
Leguat's Rail
White Mascarene Starling

Classification

EXTINCT AND ENDANGERED SPECIES

The following list comprises, in taxonomic order, those animals described in *The Doomsday Book of Animals*, and the species and subspecies listed as severely endangered in the Red Data Books of the IUCN. The Latin classifications differ in some cases from those in the Red Data Books, where our consultants have so advised.

In some cases the term 'endangered' is academic only: there is a strong likelihood, for instance, that the Kauai O-O (of which there are only 2 known specimens) will be extinct before this book goes to press; while the Kestrel and Ring-necked Parakeet of Mauritius can scarcely survive another five years. A more complex issue involves animals which may already be extinct (for example, Sclater's Lemur, the Glaucous Macaw or the Whiteline Topminnow). To avoid a proliferation of symbols, these are entered as numbering less than 100. For the same reason we have not used a separate symbol for species (such as the Arabian Oryx, several Sika deer or the Amistad Gambusia) which exist only in captivity.

Some species present special problems: the Desert Hare-Wallaby is known from only one – recent – skull and though presumably extinct could, if found, be quite numerous. It is listed as 'endangered'.

Extinct animals are indicated by capitals. The dagger (†) indicates populations of less than 100. The asterisk (*) acknowledges rumours of survival, in such cases as the Thylacine's (where there are two sharply-divided schools of thought) and the Barred-wing Rail (which *was* rediscovered but whose continued survival seems unlikely). The asterisk also attends the Society Parakeet whose extraordinary rediscovery is unverified.

KEY – the four categories used in the Classification are as follows:
BLACKFIN CISCO – extinct
River Terrapin – endangered
† Californian Condor – less than 100
* DAWSON'S CARIBOU – rumours of survival

Fishes

ACIPENSERIFORMES

Acipenseridae
Shortnose Sturgeon
Acipenser brevirostrum

SALMONIFORMES

Salmonidae
Greenback Cutthroat Trout
Salmo clarki stomias
Gila Trout
Salmo gilae
Ala Balik
Salmo platycephalus

Coregonidae
BLACKFIN CISCO
Coregonus nigripinnis
DEEPWATER CISCO
Coregonus johannae
Longjaw Cisco
Coregonus alpenae

Prototroctidae
* NEW ZEALAND GRAYLING
Prototroctes oxyrhynchus

CYPRINIFORMES

Cyprinidae
THICKTAIL CHUB
Gila crassicauda
STUMPTOOTH MINNOW
Stypodon signifer
PARRAS ROUNDNOSE MINNOW
Dionda episcopa plunctifer
BIG SPRING SPINEDACE
Lepidomeda mollispinis pratensis
PAHRANAGAT SPINEDACE
Lepidomeda altivelis
GRASS VALLEY SPECKLED DACE
Rhinichthys osculus reliquus
Cicek
Acanthorutilus handlirschi
Black Ruby Barb
Barbus (Puntius) nigrofasciatus
Fiery Redfin
Barbus phlegothon
Treur River Barb
Barbus treurensis
Border Barb
Barbus trevelyani
Mohave Tui Chub
Gila bicolor mohavensis
Owens Tui Chub
Gila bicolor snyderi
Humpback Chub
Gila cypha
Bonytail
Gila elegans
Salinas Chub
Gila modesta
Pahranagat Chub
Gila robusta jordani
Virgin River Roundtail Chub
Gila robusta seminuda
Moapa Dace
Moapa coriacea
Bluntnose Shiner
Notropis simus
Cahaba Shiner
Notropis sp.
Maluti
Oreodaimon quathlambae
Woundfin
Plagopterus argentissimus
Colorado River Squawfish
Ptychocheilus lucius
Independence Valley Speckled Dace
Rhinichthys osculus lethoporus
Miyako Tanago (Tokyo Bitterling)
Tanakia tanago
Isparta Cyprinid
Tylognathus klatti

Catostomidae
HARELIP SUCKER
Lagochila lacera
SHORTNOSE SUCKER
Chasmistes brevirostris
JUNE SUCKER
Chasmistes liorus
SPRING VALLEY SUCKER
Pantosteus sp.
Warner Sucker
Catostomus warnerensis
Cui ui
Chasmistes cujus

Cobitidae
Ayumodoki
Leptobotia curta

SILURIFORMES

Diplomystidae
Tollo de Agua Dulce
Diplomystes chilensis

Ictaluridae
Scioto Madtom
Notorus trautmani
Mexican Blindcat
Prietella phraetophila

Bagridae
Nekogigi
Coreobagrus ichikawai

Clariidae
Cave Catfish
Clarias cavernicola

CYPRINODONTIFORMES

Cyprinodontidae
LAKE TITICACA ORESTIAS
Orestias cuvieri
PARRAS PUPFISH
Cyprinodon latifasciatus
ASH MEADOWS KILLIFISH
Empetrichthys merriami
* TECOPA PUPFISH
Cyprinodon nevadensis calidae
Owens Valley Pupfish
Cyprinodon radiosus
† Santa Catarina Sabrefin
Campellolebias brucei
Leon Springs Pupfish
Cyprinodon bovinus
Devil's Hole Pupfish
Cyprinodon diabolis
Comanche Springs Pupfish
Cyprinodon elegans
Desert Pupfish
Cyprinodon macularius
Pahrump Killifish
Empetrichthys latos
† Whiteline Topminnow
Fundulus albolineatus
Caterina Pupfish
Megupsilon aporus

Poeceliidae
Amistad Gambusia
Gambusia amistadensis
Big Bend Gambusia
Gambusia gaigei
San Marcos Gambusia
Gambusia georgei
Clear Creek Gambusia
Gambusia heterochis
Gila Topminnow
Poeciliopsis occidentalis occidentalis

Monterrey
Xiphophorus couchianus

ATHERINIFORMES

Atherinidae
Key Silverside
Menidia conchorum

SCORPAENIFORMES

Cottidae
UTAH LAKE SCULPIN
Cottus echinatus
Pygmy Sculpin
Cottus pygmaeus

PERCIFORMES

Percidae
Fountain Darter
Etheostoma fonticola
Watercress Darter
Etheostoma nuchale
Okaloosa Darter
Etheostoma okaloosae
Maryland Darter
Etheostoma sellare
Snail Darter
Percina tanasi
Asprete
Romanichthys valsanicola

Sciaenidae
Totoaba
Cynoscion macdonaldi

Gobiidae
O'Opu-Alamoo
Lentipes concolor

Amphibians

CAUDATA

Ambystomatidae
Santa Cruz Long-toed Salamander
Ambystoma macrodactylum croceum

Plethodontidae
Desert Slender Salamander
Batrachoseps aridus
Texas Blind Salamander
Typhlomolge rathbuni

SALIENTIA

Discoglossidae
PALESTINE PAINTED FROG
Discoglossus nigriventer

Ranidae
VEGAS VALLEY LEOPARD FROG
Rana pipiens (onca?) fisheri

Bufonidae
Golden Toad
Bufo periglenes
Houston Toad
Bufo houstonensis

Pelobatidae
Italian Spade-foot Toad
Pelobates fuscus insubricus

Reptiles

TESTUDINES

Emydidae
River Terrapin
Batagur baska
South American Red-lined Turtle
Pseudemys ornata callirostris

Testudinidae
RODRIGUEZ GREATER TORTOISE
Geochelone (Cylindraspis) vosmaeri
RODRIGUEZ LESSER TORTOISE
Geochelone (Cylindraspis) peltastes
MAURITIAN HIGH-FRONTED TORTOISE
Geochelone (Cylindraspis) trisserata
MAURITIAN DOMED TORTOISE
Geochelone (Cylindraspis) inepta
RÉUNION TORTOISE
Geochelone (Cylindraspis) indica
RÉUNION TORTOISE
Geochelone (Cylindraspis) borbonica
MARION'S (SEYCHELLES) TORTOISES
Geochelone (Cylindraspis) sumeirei
BARRINGTON ISLAND TORTOISE
Geochelone elephantopus sp.
CHARLES ISLAND TORTOISE
Geochelone elephantopus elephantopus
ABINGDON ISLAND TORTOISE
Geochelone elephantopus abingdoni
NARBOROUGH ISLAND TORTOISE
Geochelone elephantopus phantastica
† Hood Island Tortoise
Geochelone elephantopus hoodensis
North Albemarle Tortoise
Geochelone elephantopus becki
Chatham Island Tortoise
Geochelone elephantopus chathamensis
James Island Tortoise
Geochelone elephantopus darwini
Duncan Saddleback Tortoise
Geochelone elephantopus ephippium
Vilamil Mountain Tortoise
Geochelone elephantopus guentheri
Tagus Cove Tortoise
Geochelone elephantopus microphyes
Porter's Black Tortoise
Geochelone elephantopus porteri
Cowley Mountain Tortoise
Geochelone elephantopus vandenburghi
Jervis Island Tortoise
Geochelone elephantopus wallacei
Geometric Tortoise
Psammobates geometrica

Cheloniidae
Green Turtle
Chelonia mydas
Hawksbill Turtle
Eretomochelys imbricata
Kemp's Atlantic Ridley
Lepidochelys kempii
Pacific Ridley
Lepidochelys olivacea

Dermochelyidae
Leatherback Turtle
Dermochelys coriacea

Pelomedusidae
Arrau
Podocnemis expansa

Chelidae
Short-necked Turtle
Pseudemydura umbrina

CROCODYLIA

Alligatoridae
Chinese Alligator
Alligator sinensis
Spectacled Caiman
Caiman crocodilus crocodilus
Rio Apaporis Caiman
Caiman crocodilus apaporiensis
Magdelena Caiman
Caiman crocodilus fuscus
Paraguay (Yacaré) Caiman
Caiman crocodilus yacare
Broad-nosed Caiman
Caiman latirostris
Black Caiman
Melanosuchus niger

Crocodylidae
American Crocodile
Crocodylus acutus
African Slender-snouted Crocodile
Crocodylus cataphractus
Orinoco Crocodile
Crocodylus intermedius
Morelet's Crocodile
Crocodylus moreletii
Mugger
Crocodylus palustris palustris
Cuban Crocodile
Crocodylus rhombifer
Siamese Crocodile
Crocodylus siamensis
West African Dwarf Crocodile
Osteolaemus tetraspis
False Gavial
Tomistoma schlegelii

Gavialidae
Gharial
Gavialis gangeticus

SQUAMATA

Gekkonidae
RODRIGUEZ DAY GECKO
Phelsuma edwardnewtoni
RODRIGUEZ NIGHT GECKO
Phelsuma gigas

Iguanidae
JAMAICAN IGUANA
Cyclura collei
NAVASSA IGUANA
Cyclura cornuta onchiopsis
NAVASSA ISLAND LIZARD
Leiocephalus eremitus
MARTINIQUE LIZARD
Leiocephalus herminieri
† Giant Anole
Anolis roosevelti
† Fiji Banded Iguana
Brachylophus fasciatus
San Joaquín Leopard Lizard
Crotaphytus wislizenii silus

Teiidae
MARTINIQUE GIANT AMEIVA
Ameiva major
GRAND ISLET (GUADELOUPE) AMEIVA
Ameiva cineracea
St Croix Ground Lizard
Ameiva polops

Anguidae
JAMAICAN GIANT GALLIWASP
Diploglossus occiduus

Lacertidae
RATAS ISLAND LIZARD
Podarcis lilfordi rodriquezi
SAN STEPHANO LIZARD
Podarcis sicula sanctistephani
† Hierro Giant Lizard
Gelotia simonyi

Scincidae
MAURITIAN GIANT SKINK
Didosaurus mauritianus
RÉUNION SKINK
Gongylomorphus bojerii borbonica
CAPE VERDE GIANT SKINK
Macroscincus coctei

SERPENTES

Boidae
ROUND ISLAND BOA
Bolyeria multocarinata
† Keel-scaled Boa
Casarea dussumieri

Colubridae
JAMAICA TREE SNAKE
Alsophis ater
ST CROIX TREE SNAKE
Alsophis sanctícrucis
MARTINIQUE RACER
Dromicus cursor
ST LUCIA RACER
Dromicus ornatus
San Francisco Garter Snake
Thamnophis sirtalis tetrataenia

Elapidae
Central Asian Cobra
Naja oxiana

Birds

STRUTHIORNIFORMES

Struthiornidae
* ARABIAN OSTRICH
Struthio camelus syriacus

RHEIFORMES

Rheidae
Puna Rhea
Pterocnemia pennata tarapacensis

CASUARIIFORMES

Dromaiidae
TASMANIAN EMU
Dromaius novaehollandiae diemenensis
DWARF EMU
Dromaius novaehollandiae diemenianus

DINORNITHIFORMES

Dinornithidae
MOAS
Dinornis maximus
Dinornis robustus
Dinornis gazella
Pachyornis septentrionalis
Pachyornis elephantopus
Pachyornis mappini
Emeus huttoni
Euryapteryx gravis
Euryapteryx geranoides
Anomalopteryx parvus
Anomalopteryx didiformes
Anomalopteryx oweni
* *Megalapteryx didinus*
Megalapteryx hectori
Megalapteryx benhami

AEPYORNITHIFORMES

Aepyornithidae
ELEPHANT BIRD
Aepyornis maximus

TINAMIFORMES

Tinamidae
Pernambuco Solitary Tinamou
Tinamus solitarius pernambucensis

PODICIPEDIFORMES

Podicipedidae
† Colombian Grebe
Podiceps andinus
Atitlan Grebe
Podilymbus gigas

PROCELLARIIFORMES

Diomedeidae
Short-tailed Albatross
Diomedea albatrus

Procellariidae
Black Petrel
Procellaria parkinsoni
† Réunion Petrel
Pterodroma aterrima
† Cahow
Pterodroma cahow
Cook's Petrel
Pterodroma cookii cookii
Chatham Island Petrel
Pterodroma hypoleuca axillaris
† Chatham Island Taiko
Pterodroma magentae
Galapagos Dark-rumped Petrel
Pterodroma phaeopygia phaeopygia
Hawaiian Dark-rumped Petrel
Pterodroma phaeopygia sandwichensis

Hydrobatidae
GUADALUPE STORM PETREL
Oceanodroma macrodactyla

PELECANIFORMES

Sulidae
Abbott's Booby
Sula abbotti

Phalacrocoracidae
SPECTACLED CORMORANT
Phalacrocorax perspicillatus

CICONIIFORMES

Ardeidae
BONIN NIGHT HERON
Nycticorax caledonicus crassirostris

Ciconiidae
Japanese White Stork
Ciconia ciconia boyciana

Threskiornithidae
† Japanese Crested Ibis
Nipponia nippon
Waldrapp
Geronticus eremita

ANSERIFORMES

Anatidae
COUES' GADWALL
Anas strepera couesi
† Marianas Mallard
Anas oustaleti
KOREAN CRESTED SHELDUCK
Tadorna (Pseudotadorna) cristata
PINK-HEADED DUCK
Rhodonessa caryophyllacea
LABRADOR DUCK
Camptorhynchus labradorius
AUCKLAND ISLAND MERGANSER
Mergus australis

FALCONIFORMES

Cathartidae
PAINTED VULTURE
Sarcorhamphus sacra
† Californian Condor
Gymnogyps californianus

Accipitridae
† Grenada Hook-billed Kite
Chondrohierax uncinatus mirus
† Madagascar Fish Eagle
Haliaeetus vociferoides
* MADAGASCAR SERPENT EAGLE
Eutriorchis astur
† Anjouan Island Sparrow Hawk
Accipiter francesii pusillus
Spanish Imperial Eagle
Aquila heliaca adalberti

Falconidae
GUADALUPE CARACARA (QUELILI)
Polyborus lutosus
Seychelles Kestrel
Falco araea
† Mauritius Kestrel
Falco punctatus

GALLIFORMES

Cracidae
† Red-billed Curassow
Crax blumenbachii
† Eastern Razor-billed Curassow
Crax mitu mitu
† Utila Chachalaca
Ortalis vetula deschauenseei
White-winged Guan
Penelope albipennis
† Cauca Guan
Penelope perspicax
Black-fronted Piping Guan
Aburria jacutinga
† Trinidad Piping Guan
Pipile pipile pipile
Horned Guan
Oreophasis derbianus

Tetraonidae
Cantabrian Capercaillie
Tetrao urogallus cantabricus
HEATH HEN
Tympanuchus cupido cupido

Phasianidae
Masked Bobwhite Quail
Colinus virginianus ridgwayi
NEW ZEALAND QUAIL
Coturnix novaezelandiae
HIMALAYAN MOUNTAIN QUAIL
Ophrysia superciliosa
Gorgetted Wood-Quail
Odontophorus strophium
Italian Grey Partridge
Perdix perdix italica
Cabot's Tragopan
Tragopan caboti
Western Tragopan
Tragopan melanocephalus
Chinese Monal
Lophophorus lhuysii
Brown-eared Pheasant
Crossoptilon mantchuricum
Elliot's Pheasant
Syrmaticus ellioti
Cheer Pheasant
Catreus wallichii

GRUIFORMES

Gruidae
† Mississippi Sandhill Crane
Grus canadensis pulla
Siberian White Crane
Grus leucogeranus

Rallidae
† Auckland Island Rail
Rallus (Dryolimnas) pectoralis muelleri
Light-footed Clapper Rail
Rallus longirostris levipes
WAKE ISLAND RAIL
Gallirallus wakensis
TAHITI RAIL
Gallirallus ecaudata
MODEST RAIL
Gallirallus modestus
DIEFFENBACH'S RAIL
Gallirallus dieffenbachii
MACQUARIE ISLAND BANDED RAIL
Gallirallus philippensis macquariensis
† Lord Howe Island Woodhen
Gallirallus (Tricholimnas) sylvestris
JAMAICAN WOOD RAIL
Aramides concolor concolor
MAURITIAN RED RAIL
Aphanapteryx bonasia
LEGUAT'S RAIL
Aphanapteryx leguati
† Lefresnaye's Rail
Tricholimnas lafresnayanus
LAYSAN RAIL
Porzana palmeri
SANDWICH RAIL
Porzana sandwichensis
KITTLITZ'S RAIL
Porzana monasa
LITTLE ST HELENA RAIL
Porzana astrictocarpus
* BARRED-WING RAIL
Nesoclopeus poeciloptera
IWO JIMA RAIL
Poliolimnas cinereus brevipes
WHITE GALLINULE
Porphyrio (Notornis) albus
Takahe
Porphyrio (Notornis) mantelli
SAMOAN WOOD RAIL
Pareudiastes pacificus
TRISTAN GALLINULE
Gallinula nesiotis
Hawaiian Gallinule
Gallinula chloropus sandvicensis

Rhynochetidae
Kagu
Rhynochetos jubatus

Otidae
Great Indian Bustard
Choriotis nigriceps

CHARADRIIFORMES

Haematopodidae
† Chatham Island Oystercatcher
Haematopus chathamensis
† Canarian Black Oystercatcher
Haematopus moqumi meadwaldoi

Charadriidae
New Zealand Shore Plover
Thinornis novae-seelandiae

Scolopacidae
* ESKIMO CURLEW
Numenius borealis
TAHITIAN SANDPIPER
Prosobonia leucoptera
MOOREAN SANDPIPER (?)
Prosobonia ellisi

Recurvirostridae
† Black Stilt
Himantopus novaezelandiae

Glareolidae
JERDON'S COURSER
Rhinoptilus bitorquatus

Laridae
Californian Least Tern
Sterna albifrons browni

Alcidae
GREAT AUK
Alca (Pinguinus) impennis

COLUMBIFORMES

Raphidae
DODO
Raphus cucullatus
RODRIGUEZ SOLITAIRE
Pezohaps solitarius
RÉUNION SOLITAIRE
Ornithaptera solitarius
WHITE DODO
Victoriornis imperialis

Columbidae
BONIN WOOD PIGEON
Columba versicolor
LORD HOWE ISLAND PIGEON
Columba vitiensis godmanae
† Puerto Rico Plain Pigeon
Columba inornata wetmorei
Laurel Pigeon
Columba junoniae
† Seychelles Turtle Dove
Streptopelia picturata rostrata
PASSENGER PIGEON
Ectopistes migratorius
TANNA DOVE
Gallicolumba ferruginea
CHOISEUL CRESTED PIGEON
Microgoura meeki
PIGEON HOLLANDAISE
Alectroenas nitidissima
Palau Nicobar Pigeon
Caloenas nicobarica pelewensis
† Mauritius Pink Pigeon
Nesoenas mayeri
Marquesas Pigeon
Ducula galatea
Truk Micronesian Pigeon
Ducula oceanica teroaki
NORFOLK ISLAND PIGEON
Hemiphaga novaeseelandiae spadicea
† Chatham Island Pigeon
Hemiphaga novaeseelandiae chathamensis

PSITTACIFORMES

Psittacidae
† Kakapo
Strigops habroptilus
NORFOLK ISLAND KAKA
Nestor productus
* NEW CALEDONIAN LORIKEET
Charmosyna (Vini) diadema
† Glaucous Macaw
Anodorhychus glaucus
Lear's Macaw
Anodorhynchus leari
YELLOW-HEADED MACAW
Ara gossei
GREEN AND YELLOW MACAW
Ara erythrocephala
DOMINICAN MACAW
Ara atwoodi
CUBAN RED MACAW
Ara tricolor
LABAT'S CONURE
Aratinga labati
PUERTO RICAN CONURE
Aratinga chloroptera maugei
Maroon-fronted Parrot
Rhynchopsitta pachyrhyncha terrisi
WESTERN CAROLINA PARAKEET
Conuropsis carolinensis ludoviciana
EASTERN CAROLINA PARAKEET
Conuropsis carolinensis carolinensis
GUADELOUPE AMAZON
Amazona violacea
MARTINIQUE AMAZON
Amazona martinica
CULEBRA ISLAND AMAZON
Amazona vittata gracileps
† Puerto Rican Amazon
Amazona vittata
† St Lucia Amazon
Amazona versicolor
Jacquot
Amazona arausiaca
Red-tailed Amazon
Amazona braziliensis
St Vincent Amazon
Amazona guildingii
Imperial Parrot
Amazona imperialis
† Seychelles Lesser Vasa Parrot
Coracopsis nigra barklyi
BROAD-BILLED PARROT
Lophopsittacus mauritanicus
RODRIGUEZ PARROT
Necropsittacus rodericanus
MASCARENE PARROT
Mascarinus mascarinus
SEYCHELLES PARAKEET
Psittacula wardi
† Mauritian Ring-necked Parakeet
Psittacula echo
RÉUNION RING-NECKED PARAKEET
Psittacula eques
RODRIGUEZ RING-NECKED PARAKEET
Psittacula exsul
† Beautiful (Paradise) Parrot
Psephotus pulcherrimus
BLACK-FRONTED PARAKEET
Cyanoramphus zealandicus
* Society Parakeet
Cyanoramphus ulietanus
† Orange-fronted Parakeet
Cyanoramphus malherbi
† Forbes's Parakeet
Cyanoramphus auriceps forbesi
MACQUARIE ISLAND PARAKEET
Cyanoramphus novaezelandiae erythrotis
RED-FRONTED PARAKEET
Cyanoramphus novaezelandiae subflavescens
† Norfolk Island Parakeet
Cyanoramphus novaezelandiae cookii
† Western Ground Parrot
Pezoporus wallicus flaviventris

CUCULIFORMES

Cuculidae
DELALANDE'S COUCAL
Coua (Cochlothraustes) delalandei

STRIGIFORMES

Tytonidae
MAURITIAN BARN OWL
Tyto sauzieri
NEWTON'S BARN OWL
Tyto newtoni
† Soumagne's Owl
Tyto soumagnei

Strigidae
COMORO SCOPS OWL
Otus rutilus capnodes
† Seychelles Bare-legged Scops Owl
Otus insularis
COMMERSON'S SCOPS OWL
Scops (Otus) commersoni
† Lanyu Scops Owl
Otus elegans botelensis
* SOUTH ISLAND LAUGHING OWL
Sceloglaux albifacies albifacies
NORTH ISLAND LAUGHING OWL
Sceloglaux albifacies rubifacies
FOREST SPOTTED OWLET
Athene blewitti
RODRIGUEZ LITTLE OWL
Athene murivora
ANTIGUA BURROWING OWL
Speotyto cunicularia amaura
GUADELOUPE BURROWING OWL
Speotyto cunicularia guadeloupensis

CAPRIMULGIFORMES

Caprimulgidae
JAMAICAN PAURAQUÉ
Siphonorhis americanus
† Least Pauraqué
Siphonorhis (Microsiphonorhis) brewsteri

APODIFORMES

Trochilidae
Chilean Woodstar
Eulidia yarrellii
Hook-billed Hermit
Glaucis dohrnii
Klabin Farm Long-tailed Hermit
Phaethornis margarettae
Black Barbthroat
Threnetes grzimelii

CORACIIFORMES

Alcedinidae
RYUKYU (MIYAKO) KINGFISHER
Halcyon miyakoensis
Guam Micronesian Kingfisher
Halcyon cinnamomina cinnamomina

PICIFORMES

Picidae
† Tristram's Woodpecker
Dryocopus javensis richardsi
† Okinawa Woodpecker
Sapheopipo noguchii
† Cuban Ivory-billed Woodpecker
Campephilus principalis bairdii
* IVORY-BILLED WOODPECKER
Campephilus principalis principalis
† Imperial Woodpecker
Campephilus imperialis
GUADALUPE FLICKER
Colaptes cafer rufipileus

PASSERIFORMES

Formicariidae
† Black-hooded Ant Wren
Myrmotherula erythronotos
Fringe-backed Fire Eye
Pyriglena atra

Acanthisittidae (Xenicidae)
STEPHEN ISLAND WREN
Xenicus (Traversia) lyalli
NORTH ISLAND BUSH WREN
Xenicus longipes stokesi
STEAD'S BUSH WREN
Xenicus longipes variabilis

Atrichornithidae
† Noisy Scrub Bird
Atrichornis clamosus

Laniidae
San Clemente Loggerhead Shrike
Lanius ludovicanus mearsi
Black-capped Bush Shrike
Maloconotus alius

Vangidae
† Van Dam's Vanga
Xenopirostris damii
† Pollen's Vanga
Xenopirostris polleni

Troglodytidae
GUADALUPE BEWICK'S WREN
Thryomanes bewickii brevicauda
† Guadeloupe House Wren
Troglodytes aëdon (musculus) guadeloupensis
MARTINIQUE HOUSE WREN
Troglodytes aëdon (musculus) martinicensis
ST LUCIA HOUSE WREN
Troglodytes aëdon (musculus) mesoleucus

Mimidae
† Martinique Trembler
Cinclocerthia ruficauda gutturalis
† Martinique White-breasted Thrasher
Ramphocinlus brachyurus brachyurus
St Lucia White-breasted Thrasher
Ramphocinclus brachyurus sanctae-luciae

Turdidae
† Southern Ryukyu Robin
Erithacus (Luscinia) komadori subrufa
Dappled Mountain Robin
Modulatrix orostruthus orostruthus
† Seychelles Magpie Robin
Copsychus seychellarum
BONIN THRUSH
Zoothera terrestris
BAY THRUSH
Turdus (Zoothera) ulietensis
LANAI OMAO
Phaeornis obscurus lanaiensis
OAHU OMAO
Phaeornis obscurus oahensis
MOLOKAI OMAO
Phaeornis obscurus rutha
† Kauai Omao
Phaeornis obscurus myadestina
† Puaiohi
Phaeornis palmeri
LORD HOWE ISLAND BLACKBIRD
Turdus poliocephalus vinitinctus
† Grey-headed Blackbird
Turdus poliocephalus poliocephalus

Sylviidae
† Long-legged Warbler
Trichocichla rufa
† Codfish Island Fernbird
Bowdleria punctata wilsoni
CHATHAM ISLAND FERNBIRD
Bowdleria rufescens
LAYSAN MILLERBIRD
Acrocephalus familiaris familiaris
Eiao Polynesian Warbler
Acrocephalus caffra aquilonis
† Moorean Polynesian Warbler
Acrocephalus caffra longirostris
† Rodriguez Brush Warbler
Bebrornis rodericanus

Acanthizidae
† Western Rufous Bristlebird
Dasyornis broadbenti littoralis
LORD HOWE ISLAND FLYCATCHER
Gerygone igata (ignata) insularis

Muscicapidae
TONGA TABU TAHITI FLYCATCHER
Pomarea nigra atra
Tahiti Flycatcher
Pomarea nigra nigra
Hivoa Flycatcher
Pomarea mendozae mendozae
† Nukuhiva Flycatcher
Pomarea mendozae nukuhivae
LORD HOWE ISLAND FANTAIL
Rhipidura fuliginosa cervina
† Black Robin
Petroica traversi
† Seychelles Black Paradise Flycatcher
Terpsiphone corvina

Zosteropidae
LORD HOWE ISLAND WHITE EYE
Zosterops strenua
MARIANNE SEYCHELLES WHITE EYE
Zosterops semiflava
† White-breasted Silver Eye
Zosterops albogularis
† Gizo White Eye
Zosterops luteirostris
Truk Great White Eye
Rukia ruki

Meliphagidae
KIOEA
Chaetoptila angustipluma
HAWAIIAN O-O
Moho nobilis
OAHU O-O
Moho apicalis
MOLOKAI O-O
Moho bishopi
† Kauai O-O
Moho braccatus
† Mukojima Bonin Honeyeater
Apalopteron familiare
† Helmeted Honeyeater
Meliphaga melanops cassidix
CHATHAM ISLAND BELLBIRD
Anthornis melanocaphalus

Emberizidae
ST KITTS PUERTO RICAN BULLFINCH
Loxigilla portoricensis grandis
Puerto Rican Bullfinch
Loxigilla portoricensis portoricensis
CLOUDED GALAPAGOS FINCH
Geospiza nebulosa
GUADALUPE RUFOUS-SIDED TOWHEE
Pipilo erythrophthalmus consobrinus

Thraupidae
† Cherry-throated Tanager
Nemosia rourei

Parulidae
† Bachman's Warbler
Vermivora bachmanii
Kirtland's Warbler
Dendroica kirtlandii
† Semper's Warbler
Leucopeza semperi

Drepanididae
GREAT AMAKIHI
Loxops (Viridonia) sagittirostris
MOLOKAI ALAUWAHIO
Loxops (Paroreomyza) maculata flammea
LANAI ALAUWAHIO
Loxops (Paroreomyza) maculata montana
Kauai Alauwahio
Loxops (Paroreomyza) maculata bairdi
† Oahu Alauwahio
Loxops (Paroreomyza) maculata maculata
OAHU AKEPA
Loxops coccinea rufa
† Maui Akepa
Loxops coccinea ochracea
HAWAIIAN AKIOLOA
Hemignathus obscurus obscurus
LANAI AKIOLOA
Hemignathus obscurus lanaiensis
OAHU AKIOLOA
Hemignathus obscurus ellisianus (lichtesteinii)
KAUAI AKIOLOA
Hemignathus obscurus procerus
OAHU NUKUPUU
Hemignathus lucidus lucidus
KAUAI NUKUPUU
Hemignathus lucidus hanapepe
MAUI NUKUPUU
Hemignathus lucidus affinis
GREATER KOA FINCH
Psittirostra palmeri
LESSER KOA FINCH
Psittirostra flaviceps
KONA FINCH
Psittirostra kona
Palila
Psittirostra bailleui
Ou
Psittirostra psittacea
LAYSAN APAPANE
Himatione sanguinea freethii
ULA-AI-HAWANE
Ciridops anna
MAMO
Drepanis pacifica
BLACK MAMO
Drepanis funerea

Fringillidae
SÃO THOMÉ GROSBEAK
Neospiza concolor
São Miguel Bullfinch
Pyrrhula pyrrhula murina
Red Siskin
Spinus (Carduelis) cucullatus

Ploceidae
RÉUNION FODY
Foudia madagascariensis bruante
† Mauritius Fody
Foudia rubra
Rodriguez Fody
Foudia flavicans

Sturnidae
KUSAIE MOUNTAIN STARLING
Aplonis corvina
MYSTERIOUS STARLING
Aplonis mavornata
LORD HOWE ISLAND STARLING
Aplonis fuscus hullianus
BOURBON (RÉUNION) CRESTED STARLING
Fregilupus varius
WHITE MASCARENE STARLING
Necropsar leguati
RODRIGUEZ STARLING (?)
Necropsar rodricanus
Rothschild's Grackle
Leucopsar rothschildi

Callaeidae
* HUIA
Heteralocha acutirostris
† South Island Kokako
Callaeus cinerea cinerea
North Island Kokako
Callaeus cinerea wilsoni

Cracticidae
† Lord Howe Island Currawong
Strepera graculina crissalis

Corvidae
† Marianas Crow
Corvus kubaryi
† Hawaiian Crow
Corvus tropicus

Mammals

MARSUPIALIA

Macropodidae
TOOLACHE WALLABY
Wallabia greyi
Bridle nail-tailed Wallaby
Onychogalea fraenata
EASTERN HARE-WALLABY
Lagorchestes leporoides
Desert Hare-Wallaby
Lagorchestes asomatus
GILBERT'S POTOROO
Potorous gilberti
BROAD-FACED POTOROO
Potorous platyops
ST FRANCIS ISLAND POTOROO
Potorous sp

Petauridae
Leadbeater's Opossum
Gymnobelideus leadbeateri

Peramelidae
EASTERN BARRED BANDICOOT
Perameles (bougainville?) fasciata
WESTERN BARRED BANDICOOT
Perameles (bougainville?) myosura
PIG-FOOTED BANDICOOT
Chaeropus ecaudatus
GREAT RABBIT-BANDICOOT
Macrotis lagotis grandis
Lesser Rabbit-Bandicoot
Macrotis lagotis lagotis

Dasyuridae
Kultar
Antechinomys laniger

Thylacinidae
* THYLACINE
Thylacinus cynocephalus

INSECTIVORA

Soricidae
CHRISTMAS ISLAND MUSK SHREW
Crocidura fuliginosa trichura

Nesophontidae
PUERTO RICAN NESOPHONT
Nesophontes edithae
CUBAN NESOPHONT
Nesophontes micrus
CUBAN LONG-NOSED NESOPHONT
Nesophontes longirostris
HISPANIOLAN NESOPHONT
Nesophontes paramicrus
LESSER HISPANIOLAN NESOPHONT
Nesophontes hypomicrus
LEAST HISPANIOLAN NESOPHONT
Nesophontes zamicrus

Solenodontidae
Haitian Solenodon
Solenodon paradoxus

CHIROPTERA

Phyllostomatidae
JAMAICAN LONG-TAILED BAT
Phyllonycteris aphylla
PUERTO RICAN LONG-TONGUED BAT
Phyllonycteris major
HAITIAN LONG-TONGUED BAT
Phyllonocteris obtusa
LESSER FALCATE-WINGED BAT
Phyllops vetus
PUERTO RICAN LONG-NOSED BAT
Monophyllus plethodon frater

Natalidae
CUBAN YELLOW BAT
Natalus major primus

Vespertilionidae
Grey Bat
Myotis grisescens

Megadermatidae
Ghost Bat
Macroderma gigas

Hipposideridae
Singapore Roundleaf Horseshoe Bat
Hipposideros ridleyi

Pteropidae
† Rodriguez Flying Fox
Pteropus rodricensis
† Guam Flying Fox
Pteropus tokudae

PRIMATES

Lemuridae
Black Lemur
Lemur macaco macaco
Red-fronted Lemur
Lemur macaco rufus
Sclater's Lemur
Lemur macaco flavifrons
Sanford's Lemur
Lemur macaco sanfordi
Red-tailed Sportive Lemur
Lepilemur mustelinus ruficaudatus
White-footed Sportive Lemur
Lepilemur mustelinus leucopus
† Hairy-eared Mouse Lemur
Allocebus trichotis

Indriidae
Indri
Indri indri
Verreaux's Sifaka
Propithecus verreauxi

Tarsiidae
Philippine Tarsier
Tarsius syrichta

Callitrichidae
Golden Lion Marmoset
Leontopithecus rosalia
Buff-headed Marmoset
Callithrix flaviceps
Pinche
Saguinus oedipus oedipus

Cebidae
Red-backed Squirrel Monkey
Saimiri oerstedi
Yellow-tailed Woolly Monkey
Lagothrix flavicauda
Woolly Spider Monkey
Brachyteles arachnoides

Cercopithecidae
Drill
Mandrillus leucophaeus
Tana River Mangabey
Cercocebus galeritus galeritus
Lion-tailed Macaque
Macaca silenus
Tana River Red Colobus
Colobus badius rufomitratus
Preuss's Red Colobus
Colobus badius preussi
Pig-tailed Langur
Nasalis concolor
Douc Langur
Pygathrix nemaeus

Pongidae
Pileated Gibbon
Hylobates pileatus
Silvery Gibbon
Hylobates moloch
Orangutan
Pongo pygmaeus
Mountain Gorilla
Gorilla gorilla beringei

LAGOMORPHA

Leporidae
Ryukyu Rabbit
Pentalagus furnessi
Volcano Rabbit
Romerolagus diazi
Hispid Hare
Caprolagus hispidus

RODENTIA

Sciuridae
Delmarva Peninsula Fox Squirrel
Sciurus niger cinereus

Muridae
CAPTAIN MACLEAR'S RAT
Rattus macleari
BULLDOG RAT
Rattus nativitatis
SOUTH AUSTRALIAN SPINY-HAIRED RAT
Rattus culmorum austrinus

Capromyidae
CUBAN SHORT-TAILED HUTIA
Geocapromys columbianus
CROOKED ISLAND HUTIA
Geocapromys ingrahami irrectus
GREAT ABACO HUTIA
Geocapromys ingrahami abaconis
HISPANIOLAN HEXOLOBODON
Hexolobodon phenax
HISPANIOLAN HUTIA
Plagiodontia spelaeum
Cuvier's Hutia
Plagiodontia aedium
Dominican Hutia
Plagiodontia hylaeum
PUERTO RICAN ISOLOBODON
Isolobodon portoricensis
HAITIAN (HISPANIOLAN) ISOLOBODON
Isolobodon levir
HISPANIOLAN NARROW-TOOTHED HUTIA
Aphaetraeus montanus

Echimyidae
HISPANIOLAN SPINY RAT
Brotomys voratus
LESSER HISPANIOLAN SPINY RAT
Brotomys contractus
CUBAN SPINY RAT
Boromys offella
LESSER CUBAN SPINY RAT
Boromys torrei

(Heteropsomynae)
PUERTO RICAN 'AGOUTI'
Heteropsomys insulans
LESSER PUERTO RICAN 'AGOUTI'
Homopsomys antillensis

Heteromyidae
Morro Bay Kangaroo Rat
Dipodomys heermani morroensis

Cricetidae
JAMAICAN RICE RAT
Oryzomys antillarum
ST VINCENT RICE RAT
Oryzomys victus
MARTINIQUE MUSK RAT
Megalomys desmarestii
ST LUCIA MUSK RAT
Megalomys luciae
BARBUDA MUSK RAT
Megalomys audreyae
Salt-marsh Harvest Mouse
Reithrodontomys raviventris

Heptaxadontidae
PUERTO RICAN QUEMI
Elasmodontomys obliquus
HAITIAN QUEMI
Quemisia gravis
PUERTO RICAN CAVIOMORPH
Heptaxadon bidens

CETACEA

Balaenopteridae
Blue Whale
Balaenoptera musculus
Humpback Whale
Megaptera novaeangliae

Balaendae
Bowhead Whale
Balaena mysticetus
Black Right Whale
Eubalaena glacialis

Platanistidae
Indus Dolphin
Platanista indi

CARNIVORA

Canidae
NEWFOUNDLAND WOLF
Canis lupus beothucus
KENAI WOLF
Canis lupus alces
TEXAS GREY WOLF
Canis lupus monstrabilis
NEW MEXICAN WOLF
Canis lupus mogollensis
GREAT PLAINS LOBO WOLF
Canis lupus nubilus
SOUTHERN ROCKY MOUNTAIN WOLF
Canis lupus youngi
CASCADE MOUNTAINS BROWN WOLF
Canis lupus fuscus
NORTHERN ROCKY MOUNTAIN WOLF
Canis lupus irremotus
FLORIDA BLACK WOLF
Canis rufus floridanus
TEXAS RED WOLF
Canis rufus rufus
Mississippi Red Wolf
Canis rufus gregoryi
SHAMANU
Canis (lupus?) hodophilax
WARRAH
Dusicyon australis
† Northern Simien Fox
Canis simensis simensis
Southern Simien Fox
Canis simensis citernii
† Northern Kit Fox
Vulpes velox hebes

Ursidae
ATLAS BEAR
Ursus (arctos?) crowtheri
MEXICAN GRIZZLY
Ursus arctos nelsoni
KAMCHATKAN BEAR
Ursus arctos piscator
Baluchistan Bear
Selenarctos thibetanus gedrosianus

Mustelidae
SEA MINK
Mustela macrodon
Black-footed Ferret
Mustela nigripes
Marine Otter
Lutra felina
La Plata Otter
Lutra platensis
Southern River Otter
Lutra provacax
Giant Otter
Pteronura brasiliensis
Cameroon Clawless Otter
Aonyx microdon

Viverridae
† Malabar Large-spotted Civet
Viverra megaspila civettina

Hyaenidae
Barbary Hyaena
Hyaena hyaena barbara

Felidae
Spanish Lynx
Felis pardina
Pakistan Sand Cat
Felis margarita scheffeli
† Iriomote Cat
Prionailurus iriomotensis
Eastern Cougar
Felis concolor cougar
Florida Cougar
Felis concolor coryi
BARBARY LION
Panthera leo leo
CAPE LION
Panthera leo melanochaitus
Asiatic Lion
Panthera leo persica
BALI TIGER
Panthera tigris balica
† Caspian Tiger
Panthera tigris virgata
† Javan Tiger
Panthera tigris sondaica
Indian Tiger
Panthera tigris tigris
Siberian Tiger
Panthera tigris altaica
Chinese Tiger
Panthera tigris amoyensis
Indochinese Tiger
Panthera tigris corbetti
Sumatran Tiger
Panthera tigris sumatrae
† Sinai Leopard
Panthera pardus jarvisi
† Arabian Leopard
Panthera pardus nimr
† Anatolian Leopard
Panthera pardus tulliana
† Amur Leopard
Panthera pardus orientalis
Barbary Leopard
Panthera pardus panthera
Snow Leopard
Panthera uncia
ARIZONA JAGUAR
Felis onca arizonensis
Asiatic Cheetah
Acinonyx jubatus venaticus

PINNIPEDIA

Phocidae
CARIBBEAN MONK SEAL
Monachus tropicalis
Mediterranean Monk Seal
Monachus monachus
Hawaiian Monk Seal
Monachus schauinslandi

SIRENIA

Dugongidae
* STELLER'S SEA COW
Hydrodamalis gigas

PERISSODACTYLA

Equidae
TARPAN
Equus ferus
Przewalski's Horse
Equus (ferus) przewalskii
SYRIAN ONAGER
Equus hemionus hemippus
Indian Wild Ass
Equus hemionus khur
African Wild Asses
Equus asinus
QUAGGA
Equus (burchelli?) quagga
BURCHELL'S ZEBRA
Equus burchelli burchelli
Grevy's Zebra
Equus grevyi

Tapiridae
Mountain Tapir
Tapirus pinchaque
Central American Tapir
Tapirus bairdi
Asian Tapir
Tapirus indicus

Rhinocerotidae
† Javan Rhinoceros
Rhinoceros sondaicus
Great Indian Rhinoceros
Rhinoceros unicornis
Sumatran Rhinoceros
Didermocerus sumatrensis
Northern Square-lipped Rhinoceros
Ceratotherium simum cottoni

ARTIODACTYLA

Suidae
† Pygmy Hog
Sus salvanius

Cervidae
SCHOMBURGK'S DEER
Rucervus schomburgki
Fea's Muntjac
Muntiacus feae
Persian Fallow Deer
Cervus mesopotamica
Swamp Deer
Cervus duvauceli
† Manipur Brow-antlered Deer
Cervus eldi eldi
Thailand Brow-antlered Deer
Cervus eldi siamensis
† Formosan Sika
Cervus nippon taiouanus
† Ryukyu Sika
Cervus nippon keramae
† North China Sika
Cervus nippon mandarinus
† Shansi Sika
Cervus nippon grassianus
† South China Sika
Cervus nippon kopschi
Corsican Red Deer
Cervus elaphus corsicanus
† Shou
Cervus elaphus wallichi
Barbary Deer
Cervus elaphus barbarus
Hangul
Cervus elaphus hanglu
† Yarkand Deer
Cervus elaphus yarkandensis
Bactrian Deer
Cervus elaphus bactrianus
EASTERN ELK (WAPITI)
Cervus canadensis canadensis
MERRIAM'S ELK (WAPITI)
Cervus canadensis merriami
† Cedros Island Deer
Odocoileus hemionus cerrosensis
Columbia White-tailed Deer
Odocoileus virginianus leucurus
Argentine Pampas Deer
Ozotoceros bezoarticus celer
* DAWSON'S CARIBOU
Rangifer (tarandus?) dawsoni
GREENLAND TUNDRA REINDEER
Rangifer tarandus groenlandicus

Antilocapridae
Lower California Pronghorn
Antilocapra americana peninsularis
Sonoran Pronghorn
Antilocapra americana sonoriensis

Bovidae
AUROCHS
Bos primigenius
CAUCASIAN WISENT
Bison bonasus caucasicus
EASTERN BUFFALO (BISON)
Bison bison pennsylvanicus
OREGON BUFFALO (BISON)
Bison bison oreganus
Tamaraw
Bubalus mindorensis
Lowland Anoa
Bubalus depressicornis
Mountain Anoa
Bubalus quarlesi
† Kouprey
Bos sauveli
Wild Yak
Bos grunniens mutus
Western Giant Eland
Tragelaphus derbianus derbianus
Jentink's Duiker
Cephalophus jentinki
BLUE BUCK
Hippotragus leucophaeus
Giant Sable Antelope
Hippotragus niger variani
† Arabian Oryx
Oryx leucoryx
BUBAL HARTEBEEST
Alcelaphus buselaphus buselaphus
CAPE RED HARTEBEEST
Alcelaphus caama caama
Tora Hartebeest
Alcelaphus buselaphus tora
Swayne's Hartebeest
Alcelaphus buselaphus swaynei
Zanzibar Suni
Nesotragus moschatus moschatus
Black-faced Impala
Aepyceros melampus petersi
RUFOUS GAZELLE
Gazella rufina
Rhim
Gazella subgutturosa marica
Dorcas Gazelle
Gazella dorcas
Mountain Gazelle
Gazella gazella
Cuvier's Gazelle
Gazella gazella cuvieri
Slender-horned Gazelle
Gazella leptoceros
Mhorr Gazelle
Gazella dama mhorr
† Rio de Oro Dama Gazelle
Gazella dama lozanoi
Sumatran Serow
Capricornis sumatraensis sumatraensis
Arabian Tahr
Hemitragus jayakari
PORTUGUESE IBEX
Capra (pyrenaica?) lusitanica
* PYRENEAN IBEX
Capra pyrenaica pyrenaica
Walia Ibex
Capra ibex walie
Straight-horned Markhor
Capra falconeri megaceros
BADLANDS BIGHORN SHEEP
Ovis canadensis audoboni
Mediterranean Mouflon
Ovis orientalis

Bibliography

General sources, with essential bibliographies

ALI, S. & RIPLEY, S. D. 1968–74 *Handbook of the birds of India and Pakistan*, 10 vol. London: Oxford University Press.
ALLEN, G. M. 1947 *Extinct and Vanishing Mammals of the Western Hemisphere*. Spec. Publ. Am. Comm. Internat. Wildlife Protection, **11**. New York.
ARNOLD, E. N. 1979 Indian Ocean giant tortoises: their systematics and island adaptations. *Phil. Trans. Roy. Soc. Lond.* B **286**, 127–146.
BOND, J. 1961 *Birds of the West Indies*. London: Collins. (Ontario Publishing: Don Mills.)
BROWN, L. & AMADON, D. 1968 *Eagles, Hawks and Falcons of the World*. 2 vol. London: Country Life.
BURTON, J. A. (Ed.) 1973 *Owls of the World*. London: Lowe.
DELACOUR, J. 1954 *The Waterfowl of the World*. 4 vol. London: Country Life.
FORSHAW, J. M. 1973 *Parrots of the World*. Melbourne: Landsdowne. (New York: Doubleday.)
GOODWIN, D. 1970 *Pigeons and Doves of the World*. London: British Museum (Natural History).
GREENWAY, J. C. Jr 1967 *Extinct and Vanishing Birds of the World*, second, revised edition. New York: Dover.
HACHISUKA, M. 1953 *The Dodo and Kindred Birds*. London: Witherby.
HALLIDAY, T. 1978 *Vanishing Birds: Their Natural History and Conservation*. London: Sidgwick & Jackson. (New York: Holt, Rinehart & Winston.)
HARPER, F. 1945 *Extinct and Vanishing Mammals of the Old World*. Spec. Publ. Am. Comm. Internat. Wildlife Protection, **12**. New York.
IUCN (International Union for Conservation of Nature and Natural Resources) 1956 *Derniers Refuges*. (Atlas). Brussels: Elsevier.
IUCN 1966–80. *Red Data Books* (Fish, Reptiles & Amphibians, Birds, Plants and Mammals). Continuously updated loose-leaf files, to be replaced soon by hardbound books. Now the responsibility of SCMU (Species Conservation Monitoring Unit), 219c Huntingdon Rd., Cambridge, UK.
OLIVER, W. R. B. 1955 *New Zealand Birds*. Wellington: Reed.
RIDE, W. D. L. 1970 *Guide to the Native Mammals of Australia*. London: Oxford University Press.
RIPLEY, S. D. 1977 *Rails of the World*. Toronto: Fehely.
SILVERBURG, W. S. 1969 *The Auk, the Dodo and the Oryx: vanished and vanishing creatures*. Tadworth, Surrey: World's Work.
THOMPSON, A. L. 1964 *A New Dictionary of Birds*. London: Nelson.
WALTERS, M. 1980 *The Complete Birds of the World*. London: David & Charles.
WHITEHEAD, G. K. 1972 *Deer of the World*. London: Constable. (New York: Viking Press.)
ZISWILLER, U. 1967 *Extinct and Vanishing Animals*, revised English edition, F. & P. Bunnell. London: Longmans Green.

Accounts of species, habitats, voyages etc.

ABBOTT, C. G. 1933 History of the Guadalupe Caracara. *Condor*, **35**.
AMADON, D. 1950 The Hawaiian honeycreepers (Drepaniidae). *Bull. Amer. Mus. Nat. Hist.* **95**, 157–257.
ANDREWS, C. W. 1900 *Monograph of Christmas Island*. London.
ANONYMOUS [probably Tafforet] 1726 *Relation de l'Île de Rodrigue*. [Ms. in Ministère de la Marine, Paris]. Quoted in Hachisuka, *op. cit.* and in Strickland & Melville, and Vinson, *op. cit. infra*.

ANTHONY, A. W. [Birds and mammals of Guadalupe Island]. *Proc. Calif. Acad. Sci. 4th ser.*, **14**, 277–320.
ANTHONY, H. E. 1918 Indigenous land mammals of Puerto Rico, living and extinct. *Mem. Am. Mus. Nat. Hist. New series*, **2** (2), 329–435.
ANTONIUS, O. 1938 On the . . . distribution . . . of the Recent Equidae. *Proc. Zool. Soc. Lond.* B **107**.
ARNOLD, E. N. 1980 Recently extinct reptile populations from Mauritius and Réunion, Indian Ocean. *Jour. Zool.,* London **191**, 33–47.
AUDUBON, J. J. L. & BACHMAN, J. 1854 *Quadrupeds of North America,* 3 vol. New York.
BAILEY, V. 1931 Mammals of New Mexico. *North Amer. Fauna*, **53**.
BAKER, E. C. S. 1921 *Game Birds of India, Burma and Ceylon.* London.
BEECHEY, F. W. & VIGORS, N. A. 1839 *The zoology of Captain Beechey's voyage to the Pacific . . . in the Blossom . . .1825–1828.* London.
BERGER, A. J. 1972 *Hawaiian Birdlife.* Honolulu: University Press of Hawaii.
BOWES, A. 1787–89 *A journal of a voyage from Portsmouth to New South Wales and China in the Lady Penrhyn.* [Ms. in Mitchell Library, Sydney]. In Hindwood, K. A. 1932 An historic diary. *Emu*, **32**, 17–29.
BROWN, R. 1973 Has the Thylacine really vanished? *Animals*, **15**.
BRYDEN, H. A. (Ed.) 1899 *Great and Small Game of Africa.* London.
CABRERA, A. 1911 The Subspecies of the Spanish Ibex. *Proc. Zool. Soc. Lond.*
DU BOIS, LE SIEUR. 1669 Journal. *Proc. Zool. Soc. Lond.* 1884.
DU TERTRE, J. B. 1667–71 *Histoire générale des Antilles habitées par les Francois*, 4 vol. (in 3). Paris.
FINN, F. 1915 *Wild Animals of Yesterday and To-Day.* London.
FISHER, J., SIMON, N. & VINCENT, J. (Eds) 1969 *The Red Book.* [based on IUCN data]. London: Collins.
FORSTER, J. R. 1778 *Observations made during a voyage round the world . . .* London. [German edition, slightly expanded: 1779 Berlin.]
1844 *Descriptiones Animalium . . . .* (Lichtenstein, Ed.) Berlin.
GÉRARD, J. 1861 *Lion Hunting and Sporting Life In Algeria.* (Gustave Doré, illus.) London.
GOSSE, P. H. 1847 *A naturalist's sojourn in Jamaica.* London.
GOULD, J. 1840–48 *The Birds of Australia*, 7 vol. [Supplements 1–5, 1851–69]. London.
1845–63 *The Mammals of Australia*, 3 vol. London.
GÜHLER, U. [Guelher in English pubs.] 1936 Beitrag zur Geschichte von *Cervus (Rucervus) schomburgki. Zeitschrifte fur Säugetierkunde*, **11** (1), 20–31.
GUILER, E. R. 1961 The Former Distribution and Decline of the Thylacine. *Aust. Hour. Sci.* **23** (7).
1966 In pursuit of the Thylacine. *Oryx*, August 1966.
GRANT, C. 1801 *History of Mauritius.* [Deals with all Mascarene Islands]. London.
GRIEVE, S. 1885 *The great auk, or garefowl . . . . its history, archaeology, and remains.* London.
GUNDLACH, J. 1876 *Contribución a la ornitología Cubana.* Havana.
HALL, R. L. & SHARP, H. 1978 *Wolf and Man, evolution in parallel.* New York: Academic Press.
HARMANSEN, HEEMSKERCK, MATALEIF, SCHOUTEN, SPILBERGEN, VAN DEN BROECKE, VAN DER HAGEN & VAN NECKE. 1702–06 *Receuil des Voiages qui ont servi à l'établissement . . . de la Compagnie des Indes Orientales . . .* [French trans. of Dutch Journals], 5 vol. Amsterdam.
HARPER, F. 1936, 1942 [The Painted Vulture] *Auk*, 381–392; 104.
HERBERT, T. 1638 *Some Yeares' Travels into divers parts of Asia and Afrique . . . Revised and enlarged by the Author.* London
HINDWOOD, K. A. 1938 The extinct birds of Lord Howe Island. *Austr. Mag.* **6**, 319–324.
HONEGGER, R. E. 1981 List of Amphibians and Reptiles Either Known or Thought to Have Become Extinct since 1600. *Biol. Conserv.* **19**, 141–158.
HUBBS, C. L., MILLER, R. R. & HUBBS, L. C. 1974 Hydrographic history and relict fishes of the north-central Great Basin. *Mem. Calif. Acad. Sci.* **7**, 1–259.
JOHNSGAARD, P. A. 1973 *Grouse and Quails of North America.* Lincoln: University of Nebraska Press.
JONES, F. W. 1924 *The Mammals of South Australia.* Adelaide.
KITTLITZ, F. H. VON. 1832–33 *Kupfertafeln zur Naturgeschichte,* 3 vol. Frankfurt am Main.
KURODA, N. 1925 *A contribution to the knowledge of the avifauna of the Riu Kiu Islands.* Tokyo.
LEACH, H. R., NICOLA, S. J. & BRODE, J. M. (Eds) 1976 *At the Crossroads, a report on California's*

endangered and rare fish and wildlife. Sacramento: Calif. Dept. Fish & Game.
LEGUAT, F. 1708 *Voyages et Aventures de Francois Leguat, & de ses Compagnons, en deux isles désertes des Indes Orientales . . .*, 2 vol. London. [English, 1 vol., trans: *A new Voyage to the East Indies by Francis Leguat and his companions*, 1708. London].
LOPEZ, B. H. 1978 *Of Wolves and Men*. London: Dent. (1979. New York: Scribner.)
MACFARLAND, C. G., VILLA, J. & TORO, B. 1974 The Galapagos giant tortoises (*Geochelone elephantopus*). Part 1: Status of the surviving populations. *Biol. Conserv.* **6**, 118–133.
MC DOWALL, R. M. 1976 [The New Zealand Grayling.] *Aust. Jour. Mar. & Freshwater Res.* **27** (4), 641–659.
MECH, D. 1970 *The Wolf*. New York: Doubleday.
MENDELSSOHN, H. & STEINITZ, H. 1943 A new frog from Palestine. *Copeia*, 231–233.
MILLER, G. JR. 1929 [Nesophontidae.] *Smithsonian Misc. Coll.* **81** (9), 1–17.
MILLER, R. R. 1948 The cyprinodont fishes of the Death Valley system of eastern California and southwestern Nevada. *Misc. Publ. Mus. Zool. Univ. Mich.* **68**, 107–122.
1960 Man and the changing fish fauna of the American South West. *Pap. Mich. Acad. Sci. Arts*, **46**.
NELSON, E. W. & GOLDMAN, E. A. 1933 Revision of the Jaguars. *Journ. Mamm.* **14**.
NORTH-COOMBES, A. 1971 *The Island of Rodrigues*. Port Louis, Mauritius: privately printed.
OLIVER, W. R. B. 1949 *Moas of New Zealand*. Dominion Museum Bulletin, **15**. Wellington.
POCOCK, R. I. 1913 The affinities of the Antarctic Wolf. *Proc. Zool. Soc. Lond.*, 382–393.
RADDE, G. 1893 On the present range of the European Bison in the Caucasus. *Proc. Zool. Soc. Lond.*
RIDGWAY, R. 1876 Ornithology of Guadeloupe [*sic*] Island. *Bull. U.S. Geol. Geogr. Surv. Terr.* **2** (2), 183–195.
RIPLEY, S. D. 1952 Vanishing and extinct bird species of India. *Journ. Bombay Nat. Hist. Soc.* **50** (4), 902–906.
ROTHSCHILD, W. 1907 *Extinct birds, ...* London.
SCARLET, R. J. 1974 Moa and Man in New Zealand. *Notornis*, **21** (1).
SCHLEICH, H-H. 1979 Der Kapverdische Riesenskink, *Macroscincus coctei*, eine ausgestorbene Echse? *Natur und Museum*, **109** (5), 133–138.
SCHORGER, A. W. 1955 *The Passenger Pigeon*. Madison: University of Wisconsin.
SEEBOHM, H. 1890 Birds of the Bonin Islands. *Ibis*, 95–108.
SHELDON, C. 1912 *Wilderness of the North Pacific Coast Islands*. London.
SHOEMAKER, H. W. 1915 *A Pennsylvania bison hunt*. Middleburg, Pa.
SMITH-VANIZ, W. F. 1968 *Freshwater Fishes of Alabama*. Auburn, Al.: Auburn Univ. Agr. Exp. Sta.
STEINITZ, H. 1955 Occurrence of *Discoglossus nigriventer* in Israel. *Bull. Res. Comm. Israel*, **5B** (2:B).
STEJNEGER, L. 1936 *Georg Wilhelm Steller, the pioneer of Alaskan Natural History*. Cambridge, Mass.: Harvard University Press.
STREET, P. 1961 *Vanishing animals: preserving nature's rarities*. London: Faber.
STRESEMAN, E. 1950 Birds collected during Capt. James Cook's last expedition (1776–1780). *Auk*, **67**, 66–88.
STRICKLAND, H. E. & MELVILLE, A. G. 1848 *The Dodo and its kindred . . . .* London.
SWENK, M. H. 1916 The Eskimo Curlew and its disappearance. *Ann. Rep. Smithsonian Inst. for 1915*, 325–340.
TANNER, J. T. 1942 *The ivory-billed woodpecker*. Reprint. New York: Dover.
TEMPLE, S. A. 1974 Last Chance to Save Round Island. *Wildlife*, Summer 1974.
TCHERNAVIN, V. V. 1944 A revision of the subfamily Orestiinae. *Proc. Zool. Soc. Lond.* **114**, 140–233.
TRAUTMAN, M. B. 1957 *The fishes of Ohio*. Columbia: Ohio State University Press.
US FISH & WILDLIFE SERVICE. 1940 [1939 Grizzly Bear census]. Washington.
VAN DENBURGH, J. 1914 The gigantic land tortoises of the Galápagos archipelago. *Proc. Calif. Acad. Sci.* **2**, 203–374.
VINSON, J. 1953 [Round Island Boas]. *Proc. Roy. Soc. Arts & Sci. Mauritius*. **1** (3), 253–257.
VINSON, J. & VINSON, J. M. 1969 The Saurian Fauna of the Mascarene Islands. *Maur. Inst. Bull.* **6** (4), 203–320.
WARNER, R. E. 1968 The Role of Introduced Diseases in the Extinction of the Endemic Hawaiian Avifauna. *Condor*, **70**.
WATERHOUSE, G. R. & DARWIN, C. 1839 *Zoology of the voyage of the Beagle. (Mammalia)*. London.

Index

Illustration Credits

The work of artists who illustrated this book is shown on the following pages:

Peter Hayman
Colour illustrations:
pages 66–7, 70–1, 75, 77, 79, 80–1, 84–5, 91, 94–5, 99, 100–1, 103

Tim Bramfitt
Colour illustrations:
pages 18, 23, 26, 28, 29, 33, 39, 106, 111, 113, 138–9, 141, 143, 145, 147, 253, 258–9, 261, 266–7, 258–9, 261, 266–7

Mick Loates
Black and white illustrations:
pages 25, 30, 31, 35, 40–1, 44, 49, 52, 53, 61, 64, 65, 82, 83, 86, 87, 88, 89, 92, 93, 104, 105, 114, 115, 119, 124, 125, 129, 136, 137, 148, 149
Colour illustrations:
pages 42–3, 47, 51, 54–5, 59, 63, 84–5, 116–7, 126–7, 130, 133, 135

Maurice Wilson
All colour and black and white illustrations appearing on pages 152–243

Graham Allen
Black and white illustrations:
pages 249, 250, 252, 254, 255, 256, 257, 264, 265, 269

Pat Harby
pages 273–275

page 12 Passenger Pigeon (Mary Evans Picture Library)
page 15 Buffaloes at Sunset by Gustav Doré (Mary Evans Picture Library)
page 16 Didunculus, or Little Dodo (The Mansell Collection)
page 150 The Tarpan (Peter Newark's Historical Pictures)
page 244 Detail from 'Giant Tortoises of the Galapagos Islands' (Peter Newark's Historical Pictures)

Atlas of Extinct Species page 276–7 Thames Cartographic Services
Jacket: Bali Tiger by Maurice Wilson, Rodriguez Little Owl by Mick Loates